SYSTEMATIC
CONSTRUCTION INSPECTION

SYSTEMATIC CONSTRUCTION INSPECTION

RALPH W. LIEBING
Registered Architect

Building Commissioner
Hamilton County, Ohio

Adjunct Instructor
OMI College of Applied Science
University of Cincinnati

A Wiley-Interscience Publication
JOHN WILEY & SONS, INC.   New York   Chichester   Brisbane   Toronto   Singapore

Systematic Construction Inspection has been written as a general overview of the current method of inspecting constructing activities. It is meant to reflect some of the conditions and situations that exist today, and in no way excludes other contractual configurations or lines of responsibility. It is impossible to portray every conceivable situation; this is general background material, and not technical or legal advice.

This book has used the established, standard construction vocabulary, which is replete with words of masculine gender, such as workman, manpower, foreman, and others that seem to denote men only. Since the English language does not include neuter personal pronouns, where masculine pronouns are used, they are intended to apply equally to men and women.

R.W.L.

Copyright © 1982 by John Wiley & Sons, Inc.

All rights reserved. Published simultaneously in Canada.

Reproduction or translation of any part of this work beyond that permitted by Sections 107 or 108 of the 1976 United States Copyright Act without the permission of the copyright owner is unlawful. Requests for permission or further information should be addressed to the Permissions Department, John Wiley & Sons, Inc.

ISBN 0-471-08065-9

Printed in the United States of America

10 9 8 7 6 5 4 3 2 1

*To my dear wife, Arlene, and
our wonderful children, Allen and Alissa*

To my mother and my in-laws

*To the readers: I hope this adds to your
understanding and leads to better project
relations.*

PREFACE

Debate has persisted for years over whether there are such things as accidents or whether all these unfortunate incidents can be attributed to carelessness. Many people still refer to major disasters as accidents, but accidents are direct results of carelessness. What if a building code clearly stated "no carelessness shall be permitted during the construction or occupancy of any building in this jurisdiction"? As hard as this would be to enforce, it addresses the crux of the inspection system in construction.

The inspection system is a mandatory, interdisciplinary network of communications, exchanges of information, and on-site observations of work in place and in progress to ensure compliance with approved contract documents and applicable codes and standards.

This is a "ten-dollar definition" of a much needed but perhaps impossible task. Construction inspection is not an individual function; it is a system, a coordinated effort by all people on the project site. Some inspectors are tainted by self-interest, but the final aspect of the project must be its abidance with the contract documents, standing laws, and codes. There is no turning away from this principle.

For the design professional, inspection is vital in finding discrepancies that have filtered through the contract documents. Recently some 11,000 lawsuits filed against design professionals were analyzed. The vast majority of the claims in the suits revolved around faulty contract documents. With time so pressing, design professionals are hard put to provide comprehensive, objective checking of their documents prior to the bidding and contract award. Although cost may be added to provide the missing work or to repair inadequate work, inspection can reveal the discrepancies early and eliminate major insufficiencies or disaster.

Like the design professional, the contractor carries a heavy responsibility for the proper construction of the project. Inspection can reveal installation discrepancies and again can deter disaster. The finest and most detailed investigations of materials and construction methods by the design professional are for naught if the final installations are not inspected and deemed proper. Faulty installation can negate the best qualities of any material or system. Liability and responsibility would fall back upon the contractor, in the first instance, but it also lies with the design professional, who selects the material. Inspection is absolutely necessary to ensure proper installation, despite a myriad of variables.

The various operatives on a construction site are trained very strictly in their own work. Very little cross-training is offered, and only on-the-job exposure leads to a tolerable acceptance of other workers. Design professionals respect contractors, for the most part, and vice versa. Respect builds with the experience of working together over a period of time. Great winter "war stories" are freely told about the inner workings of construction jobs, but in the final analysis good men on both sides know the other person and the merit of his effort.

When the public enters the completed structure, no one, save a government agency, is on hand to protect the public interest. Several government agencies are charged with code enforcement, and this in turn provides for public health, safety, and welfare. Like the contractor, and unlike the design professional, the government inspector does not care what is built as long as it meets the code.

This effort has been made to cross the jurisdictional lines, as well as the educational gap, between construction personnel. It is essential in these days of high cost and tremendous responsibility that everyone understand that inspection is truly a system intertwined among everyone on the site. A coordinated and cooperative effort, respecting each participant's interests, must be made. I hope these pages will bring some understanding to each participant; everyone is trying to do the best job possible, and inspection can only help.

The design professional is interested in seeing his project properly and carefully executed; he protects his interest and that of his client.

The contractor is interested in seeing that his liability is minimized and that a profit is turned; no profit has yet outstripped liability in the aftermath of a disaster.

The government inspector is there to protect the public interest, as required in the codes.

Inspection is the eye of enforcement, but also the eyes of excellence, pride, profit, compliance, safety, and life-preserving occupancy.

It is a worthy task.

RALPH W. LIEBING

Cincinnati, Ohio
October 1981

CONTENTS

    INTRODUCTION

1. **NEW CONCEPTS**   1
   - Changing Attitudes and Concepts   1
   - The Need for Better Understanding and Cooperation   2
   - The Complicated System of Communication   6

2. **THE DESIGN PROFESSIONAL**   9
   - Contractual Obligations   9
   - Chief Coordinator for the Entire Project   11
   - On-Site Inspection Techniques   16

3. **THE CONSTRUCTION MANAGER**   22
   - Guidance   22
   - Contractual Obligations   23
   - Avoiding Impediments of Work Progress   28

4. **THE OWNER**   30
   - Protection of Interests   30
   - Policing the Contract   31
   - Decision Making   33
   - Cost Control   33

5. **THE CONTRACTOR**   35
   - Contractual Obligations— The Many Things Included   35
   - Resolving Information from Others   40
   - Ensuring Good Work   42
   - Coordination and Responsibility for Subcontractors   44
   - Tempo of the Job   45

6. **THE SUBCONTRACTOR**   46
   - Contractual Obligations   46
   - Responsibility to Every Member of the Design/Build Team   46
   - Protection of His Rights, Interests, and Business   49

7. **THE BUILDING INSPECTOR**   53
   - Background   53
   - Parameters of His Job   58
   - The Need to Understand Various Construction Systems   60
   - The Authoritative Bystander   63

8. **CONCLUSION**   69

    APPENDIX   77

    INDEX   117

INTRODUCTION

INSPECT *tr. v.* —spected, specting, spects 1. To scrutinize carefully and critically, especially for flaws. 2. To review or observe officially.
INSPECTION *n.* 1. The act of inspecting. 2. Official observation or review.
INSPECTIONAL, *adj.*

Inspection—the very word seems to make the world stand still for a moment: What is going to happen to me? What is wrong with my work? Whether it is the kindergartener awaiting "clean hands" inspection or the raw military recruit anticipating that Saturday morning ordeal, the human heart jumps at the simple mention of inspection. It connotes a problem or some kind of negative experience. The definition itself seems to be directly opposed to the great American axiom of "innocent until proven guilty." It implies that there are flaws or irregularities, that some thing is not proper.

In construction the inspection system is extremely important to a great many people for varying reasons. All the participants on the project, as well as the using public, are direct beneficiaries of inspection. The design professional, the owner, the contractor, and the subcontractor all share an inspection function. All vitally need this function for their own welfare. The government inspectors (building, fire, sanitary, engineering, etc.) provide the necessary safeguards for the public, since the general public is not able to do this for itself; this is a proper function of the political jurisdiction. It is interesting to explore the various inspection functions and to see how they vary from party to party. Also, interpretations and basic definitions vary with each party, as each forms an inspection philosophy.

Other definitions from more construction-oriented sources soften the basic dictionary definition.

INSPECTION *n.* 1. Examination of work completed or in progress to determine its compliance with contract requirements. 2. Examination of the work by a public official, owner's representative, or others. 3. The process of measuring or checking materials, workmanship, or methods for conformance with quality controls, specifications, and/or standards.

The word *examination* implies a softer approach. It seems to allow for a more cursory look, rather than for a very exacting critical one. Perhaps the best way to approach construction inspection is to somehow modulate the inspection work between strict inspection and mere examination.

A certain tolerance must be employed in all inspection functions. Nothing is so absolute that it cannot be changed. Compliance with approved documents, manufacturers' instructions, and industry standards is the basic aim of the inspection system. Inspection techniques can be modified to see that this is accomplished, but this does not mean that all inspection must be extremely critical and minutely detailed. Much of the construction material is manufactured and installed with built-in safeguards and factors of safety. It is the wise inspector who knows what is acceptable and molds his techniques to be flexible within these safety limits. Examination that is too lenient is just as intolerable as inspection that is too severe.

There is no doubt that the contractors on construction sites do not really want to participate in any kind of inspection program. They feel it is an affront to their egos or an attack on their expertise. They feel sure that they can perform and do the work in the proper manner, at the proper time, and in abidance with budgetary restrictions. This may very well be true; but with so many interests involved, the inspection process must be constantly functioning on all levels. No participant can allow this responsibility to be ignored because not only his best interest but that of the entire project will suffer. There is an extremely delicate balance that must be struck by the inspection system as a whole to allow proper and consistent progress without allowing substandard or noncomplying work. Sometimes inspection can degenerate into an absolute intrusion on the project itself.

It is important that all the various inspection agencies at least understand one another's province if they cannot come to some kind of an abiding coordination and cooperation. Surely, no project or its contractor should be exposed to a sequence of inspections that countermand one another, that demand varying degrees of performance, and that really are counterproductive in the overall view of the project.

There is no doubt that the various interests represented by inspection groups should be represented and active in the project itself. In the end the contractor, subcontractors, and the individual workman can rest more easily knowing that the proper safeguards have been taken, knowing that they have produced the project in the best manner possible, and that any repercussions, from simple adjustments to major problems, can be backed up with evidence that the project was built in a prudent, cautious, proper manner.

Without the inspection sequence, no such evidence is available. It is possible that the project can be very suspect, and those who participated in the project can be held liable for the performance or, more properly, the nonperformance of the project.

Inspection can be a constructive element of the project if it is handled properly by the various agencies involved. For any participant to misunderstand his particular role and the limits of that role in the project can lead to chaos and an untenable situation. In the following pages, the roles of the various agencies are explored. Although the inspecting agencies cannot be regulated, the suggestions, the checklists, and general discussion should lead the agencies to a better understanding of one another. To know what the other people are doing, to know how far they can go and exactly what their limits are, and to know their proper input into the project can encourage all the inspection agencies to work as a team. In the sports world emphasis is often placed on team work, where one member of the team is so familiar with another that he knows before action begins exactly how his teammate will react, what he will do, what he can't do. And certainly this type of team effort on a construction projeot is not only refreshing to all the participants, but is essential to the proper and prompt construction and occupancy of the project.

Unfortunately, the construction "team" never really has time to fully develop and become coordinated. There simply is not time to practice working together as efficiently as a sports team. A construction project is built in place and the team must function as best it can while the work is continuing. We do not have the luxury of building, testing, evaluating, and adjusting (repeatedly if necessary) prior to the use or construction of the project. Once the process of construction starts, those involved must proceed with all caution and diligence. Besides, even after the extensive "trial" process, the mere use of the building can downgrade the protection that is built in.

It is essential for all participants in the construction process to understand the necessity for construction inspection. It simply must be done! It is unfortunate that companies and corporations often do things that individuals would not. Our economy is one of "buyer beware"; although this is both philosophical and legal, it becomes quite a problem in the end product. In many instances the buyer can actually oversee the process of the building as the seller is doing the work and building the project. This leads to a situation where the participants, instead of working as a team, actually can get into adversary positions.

The adversary condition can be perpetuated from previous projects that were not well run. One contractor, for one reason or another, may have a financial claim, or a lack of respect, or may have had a bad experience with another participant in the project. This is unfortunate; but it is human nature to try to do one's job as one judges it should be done, with a sense of pride and integrity. The variance of judgment can cause problems in cooperation.

No project can successfully tolerate adversary conditions. Whether this is manifested in simple pique ("I'm not talking to you") or open aggression ("I'll get there before he does and do my work—the heck with him"), the overall project climate will be highly charged. This can lead only to frayed nerves and more confrontations. The final project can be affected in a number of adverse ways—not meeting the schedule, low quality work, work stoppage, need to redo work, bad decisions, and so forth. Such a situation must be handled and remedied at the earliest possible time.

It is most important, however, that the inspection sequence be carried forward no matter what the climate of the project, so the end result will reflect the best possible coordinated effort. Whether this is done in a team atmosphere or in an adversary position, the project should be the primary goal of all participants. If it is not, the participants themselves must do some deep soul-searching and discuss problems frankly among themselves to ensure that the project does not suffer because of individual prejudices.

CHAPTER 1
NEW CONCEPTS

> If you are doing something the same way you have been doing it for ten years, the chances are you are doing it wrong.
>
> Charles Kettering

CHANGING ATTITUDES AND CONCEPTS

Since the mid-1960s, there have been more and more formalized efforts to reduce the time between initiation of a project and actual occupancy of the building. In an economy where it is increasingly costly to borrow money and the need for more immediate use of space allocations is becoming urgent, trying to reduce the project time has become, perhaps, the most vital concern of the construction industry.

Productivity, of course, is most important as far as the individual workman is concerned. He is capable of laying so many bricks per day, installing so many panes of glass per day, and so forth, for each of the trades. Productivity, however, cannot be increased over and over again, because it simply is not within human capability to become ever more efficient. Scheduling should take into account normal productivity. It is risky to schedule a project based on a greatly increased productivity per person. Workers can be added for increased total productivity, but increased costs will reflect the extra personnel. The proper time should be allowed for each operation, in any event.

This does not suggest that the construction workers should have a slovenly attitude, doing the least amount of work or just enough work to get by. The attitude should ideally be one that strongly encourages maximum productivity, that tries to achieve a truly coordinated effort of combining procedures, taking proper shortcuts, and seeking new methods to reduce overall project time.

Many different sequences and procedures have been introduced and then discarded or instituted. It makes no difference really which came first, fast-tracking or construction management. The two combined do indeed cut down overall project time.

Although only recently developed, fast-tracking and construction management have worked extremely well in tandem and separately. Large sums of money have been saved, and very pleased owners have gained occupancy months ahead of traditional time schedules.

Fast-tracking is a system whereby certain phases of the construction are actually in the process before the complete complement of construction documents is finished. For example, the foundation work may very well be going on while the architects and engineers are still working on the upper floor plans, elevations, and mechanical systems drawings for the building. While both the field workmen and the office forces are working toward the same goal, they are slightly out of phase with one another, with the field processes naturally lagging behind the drawings. This sequence allows for the collapsing of the time to produce documents, and time is not wasted. The project is not standing in limbo while the design professionals complete all the documents before any work is actually started.

Of course, certain risks are involved. New approaches must be taken to initiate fast-tracking. Obviously, if there are major changes in the building after the foundations have been designed and laid, there could be a very serious problem with either underdesigned footings or an inadequately designed superstructure. In most instances the building must be designed in an order contrary to the more traditional one. In years past the design professional would know the total loading of the building before the structure, the footings, and the foundation were designed. Today, this is not always so, and one must rely on carefully calculated guesses in regard to foundation design.

Experience and familiarity with the system have eased the pangs of change. The new method is widely accepted and easily incorporated as standard office procedure, and the benefits of the end results far outweigh the changes required of the contractors and design professionals.

Construction management is a new concept in overall control of the project. Although there are still bidding periods and various contractors on the job, the responsibility for the overall management and coordination of the project is in the hands of the construction manager, who will not have any of his work force actively involved in the actual building process.

The construction manager may have any one of various backgrounds: an architectural or engineering firm or perhaps a general contracting firm. Although formal education and degree programs in construction management are now available, the first ventures were initiated by established firms. They sought additional income and profit by offering expanded services or diversification. The design professional could maintain tighter control, and the general contractor could enter a business with far less risk and more reliable income.

Proper management of the project is essential to the end result and to the timing of the project occupancy. The manager, working in concert with both the field operations and the design professionals, must be prepared to influence both the planning and the construction phases. He must ensure that all processes progress at proper intervals and in proper sequence.

For instance, the manager may be working with the architect on the interior of the building, moving toward biddable documents for interior partitions. At the same time, he may be coordinating with the structural engineer and the contractor for the footings and foundations that are being installed in the field. This type of management effort can be achieved only by a person not actively engaged in the actual building process; this is the key to the construction manager system. The manager's role could be compared with that of the old general contractor. The general contractor was concerned not so much with the planning and the sequencing as with the actual construction of the project. The difference is a simple matter of priority and job description.

THE NEED FOR BETTER UNDERSTANDING AND COOPERATION

In the past the individual contractor had been left to his own devices. His skills and abilities were not as important to the project as they are when the project is being worked on according to the team concept. *Team* immediately suggests tighter coordination, a willingness to work together, a greater individual flexibility, and a need to sacrifice for the single effort. A brand-new atmosphere must be created under these circumstances. With the more modern approach and the effort at collapsing the time sequence, the atmosphere on the project must be studied and understood by all participants.

Many factors contribute to the project atmosphere. Among these are the actual construction documents, the plans and specifications themselves, which directly affect all participants. How practical the design might be (with regard to ease of construction) and how complete the documents are directly affect the working ability of the various participants. This also has a direct impact on the inspection cycle.

If many errors, omissions, ambiguities, overlaps, and contradictory information have been included in the documents, all participants must understand the existence of these items and work even harder as a team to solve the problems. For one contractor to go off on his own, or for one inspector in a cycle to suddenly become very overfastidious and impractical, does not serve the project well. Thinking and procedures must be understood and reexamined. The participants must come together to reevaluate this particular phase of the project.

Another aspect of project atmosphere is the ability of each and every contractor on the project. If the contractor traditionally keeps his projects very close to the letter of the contract or if he believes that he has submitted a bid that is too low—these actions or perceptions have a direct impact on the cooperative effort of the team, on the inspection cycle, and, of course, on the project atmosphere and the completed project.

It may well be that the contractor's attitude is something that will have to be dealt with by the other participants. Of course, each contractor brings his own attitudes and business approaches to the job. He should be accepted into the team at face value and should be instructed, literally, in the new approaches being taken on the job—not by grabbing him by the lapels and making him totally acquiesce to the desires of others, but by strongly attempting to orient him toward team work.

The actual capabilities of the contractors also affect the project's atmosphere. In a very tight economy where there is not a lot of proposed construction work, contractors often bid on projects for which they lack experience and many times they try to "move up" into areas of construction in which they have never been active before. Although one expects to hear the often-voiced complaint of "How do you get a job without experience, and how do you get experience without a job?," no project is a training ground for contractors or an opportunity to try something new; failure to perform well is a disaster, not a simple end result. There is no intentional effort in the industry to exclude or prohibit "new" contractors, but these contractors should be fully aware of the responsibilities and risks in such new ventures. A tremendous effort, an excellent attitude, and a willingness to learn must be forthcoming. Success, of course, can lead to widened business horizons and a brilliant career. Failure can lead to litigation, financial disaster, and working for someone else. The toughest battle to be fought involves finding an owner and a project team that will give a new contractor a chance. Projects are too dear to allow a frivolous contract award.

All phases of the team and the inspection cycle must somehow come to terms with this type of problem, as it may appear on any project. Several such problems may arise, and there must be proper reaction to them; one should not overreact and certainly one should not ignore the problem. Care must be taken not to allow the problem to overshadow the work.

---

2. Immediately prior to placement of pavement base, proof roll all subgrade sectors in the presence of Architect and Soils Engineer employed by Owner. Use standard rubber tired or flat-wheeled compaction equipment as well as loaded truck similar to type used in paving construction operations. Exercise critical visual observation to subgrade reaction, and any areas which exhibit instability or which react excessively under applied load are to be immediately reworked or undercut as directed by the Architect.

Figure 1. Excerpt from typical specification. Such broad-ranged, detailed requirements for one phase of the work may be new to some contractors. Note: language requiring inspection.

Beyond the individual contractor's capability, what is the ability of the entire construction team? Is the team balanced; highly skilled, highly motivated, and well organized? Or is this a combination of streetwise, wily contractors and young, eager, inexperienced contractors? Is there dangerous imbalance? Is there lack of skill? Are resources and experience adequate?

4 Systematic Construction Inspection

Not only does the inspection sequence have to acknowledge these conditions, but the entire team effort must be adjusted.

Finally, each region of the country has its own trade atmosphere. For some reason, certain areas of the country seem to have more problems with projects than other areas do. Trades react differently and different types of manpower and work are available, for not all kinds of work are done in all sections of the country. A project with a new materials system or an innovative design may encounter problems because of a lack of trade familiarity. In some areas buildings are almost exclusively steel framed, for instance. To introduce a concrete frame project would demand a level of expertise and skilled manpower that the local contractors may not be able to supply. Out-of-town firms must then be attracted, which in itself can be a tough job for a single project, especially in times of heavy construction activity. Costs can increase drastically, whether other firms come in or local firms try to deal with unfamiliar systems. In either event, it is obvious the project will suffer. Again, proper reaction is required by not only the inspection sequence, but by the entire design and construction team.

Beyond the project atmosphere are the actual field conditions at the job site itself. It may well be that part of the team, particularly the design professionals, can be working in remote air-conditioned offices, well lighted, and well heated, with all the resources they need available to them. In this atmosphere they may become complacent and not accurately assess what is happening in the field. Can the work being detailed be executed properly in the field? Are there conditions, whether they be weather, noise, or other environmental factors, that may affect the manpower and, perhaps, the work itself?

The inspection team also must work within this environment and must adjust, just as all members of the construction team must take field conditions into account. There is no choice for the inspection system but to be flexible and willing to change its techniques and methods to those that are best for the particular project. Failure to adjust can jeopardize the project.

In today's marketplace, where the buyer must beware of what he buys, and with the tremendous amounts of money being spent for construction jobs, owners are demanding, more and more, that the quality of the job meet the value they place on it. The owner is not in a position to control all the variables—scope, cost, and quality—of his project. It can easily be seen that an owner is most interested in the scope and cost—that is just how much building he can possibly build for the money he has available. With scope and cost being tightly controlled by the owner, the contractors, the entire design team, and the inspection system must be extremely active in the only flexible element left in the program—quality.

There is no doubt that there is a great deal of fluctuation in the quality range on any project. Something can always be found that costs less and still meets the project requirements. However, the inspection system must monitor all substitutions to ensure that no minimum standards are violated.

Because of the time involved in any building project, the owner has a chance to oversee almost every minute detail of it. Therefore, he is cognizant of all shortcomings that may appear for one reason or another in the project. Most certainly, he will be looking to the inspection system to produce the highest quality that can possibly be had from the various elements of the project environment, and he will want to achieve the highest quality with a minimum of effort on his part. He will want excellence, and part of the job of the inspection system is to see that excellence is produced in the very best way possible given the project conditions.

Obviously, there are shortcomings in any particular project; 100% "sterling" excellence cannot be achieved in any job. The individual workman, though he may be full of pride and skill, still may not be able to produce, for varying reasons, the excellence that the owner expects. But it is up to the inspection system on that particular project to see that the project is produced in the best possible fashion. This means that each member of the inspection system must be very active in ensuring that each part of the project within his responsibility is executed properly, whether he is viewing the basic design or the "extras" that are put into the project or simply seeing that minimum standards are being met.

Many factors can vary on any project. It is not the job of the inspection system to try to force a particular fashion or method of construction or to see that excellence is produced merely by doing the same work over and over again. The inspection team must be fully aware of what level of competence can be achieved and turned over to the owner. This level of excellence will vary from project to project because of the varying conditions, previously discussed, that create the project environment.

The methods of construction, whether techniques of building or of management, old or new, still rely on the inspection system as the basic quality control element. It is important for all members of the design/construct team to understand exactly where and how each element of the inspection system fits into the job. What is the basic criterion? What are the basic areas of responsibility? Until these are understood, it could well be that members of the team will be vying for position. There might even be many open attempts to circumvent or gloss over one area of inspection or another. The inspection system must be accepted, and the flexing of ego muscles must cease, so that inspection is in place and active at several levels. All should be given proper due for knowing their jobs, being able to perform properly, and having the integrity to do their best work.

The inspection system in construction is not set up to be obtrusive or to retard progress. The process should be a continual, progressive monitoring of the construction work. If the inspection system does become obtrusive, it is being counterproductive to the project and should not be tolerated.

Neither should any activity that tends or attempts to circumvent any of the inspection system be tolerated. The system should be met head on. The system should be open to all participants so that they know exactly what is expected of them. Their work should be adjusted to meet the criteria of the various inspection agencies.

If a project, unfortunately, breaks down so that some participants in the design/construct team become adversaries (in other words, they simply don't talk to each other), the project is in deep trouble. This element of the project environment is so necessary that it must be established at the very beginning of the project, and it must be nurtured constantly so that it functions continuously and properly. It may very well be that early in the project some basic meetings will have to be held to eliminate misunderstandings. There may be a need for some simple instruction sheets to be passed out to the participants on how to communicate, so that there is no chance for misunderstanding.

Out of small misunderstandings come the big problems, the problems of adversary positions and, worse, noncommunication.

There is also a tremendous need for feedback on a construction project. If a problem has developed in the field and is not fed back through the communication system at the proper level for resolution, the problem will burgeon. The problem will produce more adversary feelings and could eventually strangle the project. This could be manifested as a work slowdown, a full strike, a loss of productive time, or another situation that could drastically affect not only scheduling but quality and budgeting.

Briefly, problems of misunderstanding and feedback should be ironed out with the following guidelines in mind:

1 Approach all situations with an open mind. Try to see how the situation looks through the eyes of the other fellow:

2 Resolve to solve the problem for the good of the project. Do not reproach, accuse, or seek revenge.

3 Anticipate anxiety in both yourself and others. Everyone has some concern about his position, and this must be overcome in solving problems. Don't deal from emotion at any time. Prevent outbursts of temper. Be tolerant of everyone's viewpoints.

4 Strive to have all situations clearly stated, properly documented, and understood by all. There are times when criticism is necessary, but it should be based on a vision of the whole picture and not be a nit-picking accusation.

5 Do not deal in envy, gossip, trivialities, prejudices, or shows of pride.

6 Concentrate on making things clear at all stages of the project.

7 Debate in a shared, side-by-side inquiry. Don't try to make deals or resolve problems behind the backs of some people. In

trying to clarify and still give a firm viewpoint, preface your remarks with the phrase, "It seems to me...."
8 Try to let bygones be bygones. Things that happened on other projects simply have no relevance to the present one. Show others that this same attitude and the need to compromise are necessary for the project. Practice the Golden Rule. Use constructive action at all times.

In dealing with others and bringing everyone to understanding and good communication, there are times when feedback is absolutely necessary. Feedback should be open, and candor should be the hallmark. Feedback should be pertinent and not a discussion of "war stories." The following list is helpful in determining the quality of feedback and ensuring that it will be useful and for the good of the project:

1 Be specific and don't deal in wide-ranging generalities.
2 Refer to current work and don't bring up past projects.
3 Be sincere and establish trust among the members of the team.
4 Plan your feedback and see how it is best to address the recipients. By the time you are in a situation that needs feedback, you should pretty well know the other people, their attitudes, and their behavior patterns.
5 Study what you will present and show the other person just how this affects you.
6 Be sure that the time is right to discuss a particular situation.
7 Although it may be very difficult, be sure that the recipient understands what you are saying.
8 Of course, the recipient must be willing and able to accept the feedback. Therefore, it is important that the feedback be given in such a manner that it will not cause him to react negatively.
9 Perhaps the most important part is to ensure that the recipient of the feedback is, indeed, able to do something if he so chooses. Without this alternative, the feedback becomes valueless because, of course, there is no solution to the problem, and it will only lead to frustration if the recipient has no particular action to take.

THE COMPLICATED SYSTEM OF COMMUNICATION

There is no other choice on the modern day construction project than to have a communication system that produces clear, concise, complete, and timely communications among participants. This is so important to the success of any project that it simply cannot be emphasized enough. The communication system must be well planned and understood by all participants. It is mandatory that they all fully participate in the communication system as much as possible. Shortcuts cannot be used, nor can partial communications or improper or incomplete distribution lists.

The complexity of modern-day construction is so awesome and intricate that if the communication system breaks down even in a small way, the project itself is in deep jeopardy.

No project is immune to the need for communication. A "tight" inner-city project requires split-second scheduling and manipulation of workers, equipment, and material. An open, wide-ranging project separates functions and manpower to a tremendous degree. Obviously, communication is the necessary common denominator of all projects.

The system of communication begins at the basic level of two workmen talking with each other and coming to an understanding about a particular piece of the work. From here through a myriad of different combinations of people, the communication builds in intensity to the point that a simple phrase or sentence of instruction given at the wrong level, in an untimely fashion, and not properly distributed to all participants can stymie the progress of the project.

A tremendous amount of time and manpower is expended on any job simply to keep the lines of communication open and operating; otherwise, there is no way to tell just where or when the project will break down. In the event of a breakdown, much money and manpower could be required not only to find the problem and to solve it, but also to reestablish the lines of communication. Of course, this all works to the detriment of the project.

The most frequent cause of error in construction is the breakdown of communication. This failure stems from many sources—the distraction of outside thoughts and worries, pride of the individual, inattention, use of words that have different meanings to different people, commands that are misinterpreted, and the inclination of others to use their version of the instructions in their concept of the project. It is very important that each person concentrate on using proper wording when speaking to others and on trying to understand the person's point of view, so that when a communication is given, it can be complete in that it is properly received and put into a useful form or action.

To prevent construction failures entirely may be impossible. Human frailty and material imperfection can lead to construction failure, but miscommunication or the lack of communication should never contribute to it. The following ideas can help ensure that communication is kept open at all times and that everyone can participate.

1. Be sure that you know the person you are dealing with well before you entrust him with important orders or directions.

2. Be extremely patient; you may have to discuss, evaluate, or even reword communications at various times.

3. Be sure that both you and your listener are paying attention to the problem at hand and that you are not wandering from the subject.

4. Do not pass on incomplete information, and do not accept incomplete information.

5. If the information you have received is incomplete, not clearly transmitted, or not understood, say so. Be sure that the situation is clarified for everyone's benefit.

6. Do not carry out an obviously foolish interpretation. Check out the information, ask for clarification, and be sure that it makes sense to you before you act on it.

7. Make sure that both you and the person you are dealing with are interpreting words and phrases of construction jargon in exactly the same way.

8. If at any time a picture or a drawing can be used to clarify the situation, be sure to use one. The old saw of "a picture is worth a thousand words" is very apt in the construction industry.

Communication is discussed throughout the other chapters, with particular emphasis on the person involved. Lines of communication in the construction field must be very direct and constantly maintained. In the many years since the beginning of the construction industry, vast amounts of money have been lost due to construction failures and time losses that can be attributed directly to a breakdown in communications. It is very elementary to say so, but all

---

2.02 COMMUNICATION WITH OWNER

A. Generally, all communications with the Owner shall be through the Architect, _____ & Associates (e.g. Shop Drawings, Requests for Payment, request for drawings of items furnished by Owner, etc.).

Figure 2. Excerpt from typical specification. This shows a definite communications procedure. Although often unwritten, it leaves no room for miscommunication.

---

those on a construction site are, indeed, human beings. The main problem is how to mold these human beings, who come together for a relatively short time, into an open, freely communicating, coordinated team. It is important, though, that this be done—not only that communications are established and understanding is maintained, but that the project is of prime importance. Although everyone involved has various allegiances and may have different alternatives and goals in mind, the inspection function must be "for the project." This sounds very trite, but no matter what the allegiance of the inspector, he has a very firm commitment and obligation; he must participate in the entire project in such a manner that he will not inhibit or negatively affect it. His participation and communication should be positive, lasting, and ongoing, so that he can contribute in a very positive way, almost daily.

A successful construction project depends greatly on attitude. It is conceivable that a team on a construction project can be made up of several individuals or firms who have had very raw, openly hostile relationships in the past. The new situation requires that each person on the project somehow give up these feelings, shed the mantle of hostility, and attempt to do the job properly. To do otherwise is flirting with certain disaster, touching all concerned.

It has been said many times that even in a national election one vote can be meaningful. Similarly, if one person on the construction site or one member of the inspection team can work in a positive manner, there is a very good chance that he can influence the entire team. They may come together and work for the good of the project.

CHAPTER 2
THE DESIGN PROFESSIONAL

DESIGN PROFESSIONAL *n.* The professional collectively responsible for the design of man's physical environment including architecture, engineering, landscape architecture, urban planning, and similar environment-related professionals.

CONTRACTUAL OBLIGATIONS

The owner, in most instances, has a rather difficult time understanding the position of the design professional and the responsibilities involved. This is understandable, since the owner is expending a large sum of money on the project and the design professional largely has direct control over these funds since he has direct control of the project itself. The owner often wants the architect to have exactly the same attitude toward the project he has. Often this becomes manifest when the owner, wanting tight, close control over construction operations, desires that the architect be constantly on the site to ensure that the work is done properly, giving the owner full value for the money spent.

The major problem with this attitude is that it puts the design professional in the position of ultimate responsibility and, hence, liability. If the design professional does not perform in the manner and with the attitude the owner wants, he has one problem. But his problems are compounded if he does, indeed, adopt the attitude of the owner.

While being the direct agent of the owner, the design professional is a moderating force, taking the desires of the owner and molding them so they can be accommodated by the construction process. Without this understanding, moderation, and accommodation, deep problems can develop.

To take a construction project and overinspect or oversupervise it can be counterproductive to the design professional. Some design professionals prefer a totalitarian approach to projects, taking the attitude that no one can be right except them. Obviously, the team attitude is not present here, and very bad relations often develop. This does not make for a smoothly running project, even though the design professional feels his liability would be greatly increased if he had taken another tack. This has been refected recently in standardized contracts and conditions of the contracts from such sources as the American Institute of Architects. Phraseology of the past such as "periodic inspections," "supervision," and the like has been discarded from these documents because such terms are both misleading to the owner and highly hazardous to the design professional. This may sound as if the design professional is given certain status by the owner, but he is either unable or unwilling to perform. However, this is not the case. There should be a very open and deep relationship among the architect, engineer, and the owner, so that the owner understands exactly what he is getting for the fees he pays the design professional.

Many times, the owner will be very chagrined to find that the design professional or his representative is not constantly on the site. The owner may also find that certain work will be in-

10 Systematic Construction Inspection

> b. Paragraph 2.2, supplement by adding the following additional subparagraph:
>
> 2.2.4.1 The Architect will endeavor to observe and to check work but omissions, failures to provide proper material and failure to perform work correctly are totally the responsibility of the Contractor. The Contractor, not the Architect, is responsible for determination that all work under his Contract as it proceeds or as completed is performed and installed in accordance with the Drawings and Specifications and governing regulations. Where laws, codes or standards require supervision or inspection of portions of the Contractor's work by an Architect, Engineer or other competent or qualified person, it is the Contractor's responsibility to furnish such supervision and/or inspection to the satisfaction of the governing authority and without cost to the Owner. Such requirements shall in no way be the responsibility of the Architect or his field representative.

Figure 3. Excerpt from typical specification. A portion of Supplementary General Conditions states the exact position of the architect. This modifies the basic, widely used *General Conditions*.

stalled, meeting the requirements of the construction documents, but in such a way that it is new or unfamiliar to the owner. Having a little knowledge of construction in general can cause the owner some anguish. His experience, usually being limited, forces him to think in limited terms. When something new appears, he may feel threatened and frustrated. The design professional must address this attitude to ensure the owner's understanding and confidence. After all, the owner hired the design professional and should place a large measure of confidence in him.

However, the owner may want the situation changed. But the owner's contract with the design professional details all aspects of this type of situation.

The design professional must, under his standard of professional ethics, perform inspections at certain intervals. These are function oriented but not necessarily time oriented. Some operations are critical and require on-site inspection. The placing of concrete, for instance, requires that the architect be on-the-job to see that the proper procedures, equipment, manpower, finishing processes, and so on are involved in the operation. This does not, however, require that the architect have a force of ten people on the job all day, overseeing every minute procedure of the concrete placement or overseeing the driving of every single nail and other such small operations.

The design professional is responsible for the overall project. Through his experience, he

> 3A-11 PLACING:
>
> a. Place no concrete except when Architect's representative is present unless this requirement is specifically waived by the Architect. Give due notice to the Architect and all Contractors affected before placing concrete. Allow adequate time for installation of all necessary parts.

Figure 4. Excerpt from typical specification. A firm statement to the contractor, allows the architect to inspect and control the operation.

knows very well that there are details of construction—some may even be shown on the contract documents—whose execution is left to the expertise of the contractor. This is not to say that the design professional walks away from any obligation or ignores the job. There is a generally accepted understanding on a construction job that the contractor is well trained and informed. Therefore, he should perform the work to the very best of his ability and within the confines of the contract documents. If there is some work installed that is not proper, the contractor knows well that he is under contractual liability to remove the faulty work and rebuild it properly to the satisfaction of the design professional and the project requirements.

There are further contractual provisions that if the work has proceeded and faulty work is covered up, the design professional has the authority to have the cover removed for inspection of the underlying work. If it is faulty, the removal is done at the contractor's cost.

There are a number of safeguards involved, and the design professional should point these out to the owner. With this understanding, the owner will not be negatively influenced by the action and decisions of the design professional.

There are other techniques that the design professional can use to gain the owner's confidence, depending on the size and scope of the project. In any case, it is important that the professional and the owner have a good understanding of what is expected of each, what will happen and what will not happen, within the terms of the contract.

CHIEF COORDINATOR FOR THE ENTIRE PROJECT

Since the mid-1960s, there has been a good deal of debate about who, indeed, is the head of the design/construct team. Traditionally, this role has fallen to the architect or, using the new term, the design professional. The design professional conceives the project and translates the owner's requirements into reality. Many have felt that the planner over the years has assumed the role of the chief of the team. Some feel that the construction manager and even the contractors are now at the head.

Although this particular debate has not been resolved, there is no debate over who is in charge of the project team. It is the design professional. By virtue of his contract with the owner, he is the direct agent and the most immediate voice of the owner. He has taken the owner's concept and program and translated them into documents that, when executed, will produce the project the owner desires. Given this situation, no one else can be the leader of the team. No one else has this unique position of intermediary between the owner and the project.

As the head of the team, the design professional has a number of responsibilities that he should carry out, and many of these are not reflected in the contract documents. There is no doubt that the design professional must make the owner thoroughly aware of what is contained in the contract documents, the scope of the work involved, and the necessary procedures that the owner must perform. Be-

---

2.2.10.1   The Architect's field representative does not have the authority to approve materials substitutions, design changes, deviations from the Drawings and Specifications or changes in the Contract amount. Should such considerations arise, authorization for them must be in writing signed by the Project Architect for those items for which the Architect has authority under the Contract or by the Owner.

Figure 5. Excerpt from typical specification. A new procedure, but a very proper one that allows the architect to maintain positive control. Decisions are made in the studied atmosphere of the office, by a high level manager, and not in the hasty, narrow-scoped field operation.

cause of his contract with the owner, this is a major responsibility of the design professional. He must ensure that the owner understands clearly, and in minute detail, what his responsibilities are, and what exactly he can expect from the design/construct team.

It may very well be that some inspection procedure, other than that set up contractually, will be attractive to the owner. Of course, the design professional should participate in the selection of any other inspection procedure and of the individuals involved. Additional serv-

---

SECTION 01220 - PROJECT MEETINGS

PART 1 - GENERAL:

1.01 DESCRIPTION:

    a. Contractor for General Construction shall schedule and administer pre-construction meeting, periodic progress meetings and specially called meetings throughout the progress of the work.

       1. Prepare agenda for meetings.

       2. Distribute written notice of each meeting four (4) days in advance of meeting date.

       3. Make physical arrangements for meetings.

       4. Preside at meetings.

       5. Record the minutes; include all significant proceedings, decisions and agreements (or disagreements) reached.

       6. Reproduce and distribute copies of minutes within three (3) days after each meeting.

          (a) To all participants in the meeting.

          (b) To all parties affected by decisions made at the meeting.

          (c) Furnish three (3) copies of minutes to Architect.

    b. Representatives of Contractors, Subcontractors and suppliers attending the meetings shall be qualified and authorized to act on behalf of the entity each represents.

    c. Architect may attend meetings to ascertain that work is expedited consistent with Contract Documents and the construction schedules.

    d. Progress meetings scheduled by Prime Contractors to coordinate and expedite work of Subcontractors and suppliers are not included in this Section.

Figure 6. Excerpt from typical specification. The procedure denoted establishes a positive system of meetings on the job, providing a very good start that should serve the project well.

ices can be offered, such as additional on-site visits, joint visits by the owner and design professional, full-time representation by the design professional, the owner, or both, or full construction management.

The design professional should, without any doubt, provide competent and experienced supervisors from his own office. They should be people who are thoroughly familiar with the construction at hand and who are also able to perform all the intricacies of construction administration. It is inappropriate for a person who has been trained and experienced in massive projects and heavy construction to try to do a single small building of light construction.

And, of course, the reverse is also true. In either case the inspector will not be comfortable with the job he has, and, as a result, it may well be that his performance will be wanting.

In the contract documents the architect will usually provide for the necessary accoutrements that the project will require: job offices, telephones, testing procedures, and the like. Thus, he has at hand all the equipment and additional personnel that may be required to serve the project.

It is essential that the design professional establish an on-site "home base." The size of the project dictates the scope of these facilities. At times, facilities shared with the contractors

---

1.24 PROGRESS MEETING:

A. The following persons or their authorized representatives shall attend initial and each progress meeting:

1. Green Township representative (as required).

2. Architect and his consultants (as required).

3. General, mechanical and electrical contractors.

4. Subcontractors who are in need of or have pertinent information.

B. Initial meeting shall be provided for general review of drawings and specifications. In addition to the above required participants, the contractor's job superintendent and the job foremen for each and every subcontractor shall attend the initial meeting.

C. Progress meetings shall be held at the direction of the Architect not more than once every two weeks nor less than once a month throughout the entire construction period. Exact day and time shall be determined by mutual agreement. Progress meetings shall be held at the site in the construction office.

D. Chairman of progress meetings; Architect or his authorized representative.

E. Minutes of progress meetings; Architect shall take minutes, prepare and distribute one copy each to the Owner, the Architect and to each organization represented at the meeting. One bound volume of all progress meetings shall be maintained by the contractor in the job office unit project completion.

Figure 7. Excerpt from typical specification. Full explanation of the progress or job meeting sets forth who is to attend, who is chairman, and disposition of agenda, minutes, and so forth.

14 Systematic Construction Inspection

are adequate. However, the design professional needs a place to review drawings, make calls, hold meetings, and process various documents.

Another important function of the design professional is to convene a conference before construction begins. He should ensure that all the various participants are represented at this meeting. Here the ground rules of the project can be set forth, even to the extent that a written manual of procedures may be produced and distributed to all present. Here, also, each of the participating firms and their on-site representatives can get to know each other, exchange pertinent information, and learn of one another's individual operations.

To set this meeting before the first spade of ground is turned is essential to getting the project off on the right foot. It shows everybody that they are an integral, important part of the project and that the design professional is, in a sense, demanding cooperation and coordination without saying so. It is important to point out that at least the design professional is interested in opening up and maintaining the lines of communication, getting people to sit down at the very outset, face to face, to begin the project. The preconstruction meeting is becoming more widely used and increasingly important. Here the entire basis and tone of the project can be set forth. Documents can be distributed, communications established, priorities and procedures reviewed, and general thoughts about the project exchanged.

Beyond this, the design professional may want to set up periodic job meetings, with everyone currently working on the project participating. It may be that he will invite people who have either just left the job or who are anticipating coming on the job. It should be an open forum, with the minutes of the meeting formalized and distributed to all for the sake of continuity, placing responsibility, and establishing a job history in each participant's file.

The design professional should establish an orderly system for all the documentation of the job. Correspondence, forms, reports, shop drawings, samples, color selections, interpretations, memorandums, decisions, and minutes of meeting should be accurately produced and properly distributed. In this way the professional

---

APPLICABLE CODES

a) Throughout this specification there is reference to materials and workmanship being in compliance with applicable codes. In this context, applicable codes is to mean: the State of Ohio Building Code, the FHA Minimum Property Standards, and any Local City Building, Zoning and Fire Zone Standards. All Codes shall be the current edition, including all amendments at the time of the building permit is issued.

b) Applicable codes shall also include all codes, standards, and testing agencies referenced in any of the codes listed above and thus made a part of those codes, i.e. NFPA Standards, ASTM, ASHRAE, Underwriters Laboratories.

Figure 8. Excerpt from typical specification. This provision sets forth the basis for all work on the project.

d) All firecode drywall shall be installed in accordance with the system used for the approved applicable fire rating tests obtained by the drywall manufacturer and accepted by the state of Ohio and all other governing agencies. Installation of fire rated drywall will not be accepted unless installed in accordance with tested assembly instructions.

walls shall have one (1) layer of 5/8" fire code drywall on both sides for a one (1) hour rating.

5. First and second floor ceilings shall have one (1) layer of 5/8" fire code drywall for a one (1) hour rating. Third floor ceiling shall be ½" fire code for 3/4 hour rating.

3.03 SCHEDULE:

See Drawings for types and locations of exterior wall insulation, partitions and ceiling assemblies and systems:

a. Typical Exterior Walls: Apply single layer foil-backed gypsum drywall over wall insulation blankets and Z-furring channels. (Note: Insulation shall extend above gypsum wallboard height or in some conditions insulation is applied only to walls; see Drawings).

b. Other Interior Partitions: Apply single layer gypsum drywall both sides (thickness indicated) to metal studs.

c. Fire-rated assemblies:

1. Two (2) hour rated wall shall comply with U.L. Design U411.

2. Two (2) hour rated ceilings shall comply with U.L. Design G503.

Figure 9. Excerpt from typical specification. The upper specification provision for drywall is vague and inadequate; it does not firmly denote what is required. The lower one states exactly what assemblies are to be used; showing all the detail required for installation and providing an excellent basis for inspection.

can keep the owner, the various construction inspectors, contractors, subcontractors, and consultants fully informed about all aspects of the job.

Each participant is interested in keeping a complete job file on the project for his own use. Of course, this file in every case should be as complete and accurate as possible.

It is certainly no waste of paper to send pieces of documentation to a participant who is not directly involved with the memorandum. By virtue of this "nice to know" information, one can maintain a feel for the project and see how things are running.

One of the design professional's principal jobs on the site is to check on the contractors' schedule of values and in-place work. It is, of course, the owner's contractual responsibility to pay the invoices of the contractors. This payment should, however, be made only over the certification of the design professional and his on-site representative.

The design professional, by contract, is required to police the owner–contractor contract. He must be fully aware of the in-place work before he signs a certificate of payment. This in effect requires and obligates the owner to pay the contractors for work to date. Neither overpayment nor underpayment is proper.

As discussed earlier, on-site observations of the work are critical to its proper progress. Perhaps the key phrase is "timely observation of the work" rather than periodic observation. It is important that the design professional be represented on the job, in the right place, at the right time. To merely hide from office obligations by being on the job ensures nothing so far as the project is concerned. The ever-present "shadow" of the design professional can be irritating or even counterproductive.

The timely observations are necessary so that a proper standard of acceptability can be established. By being on the job at the right time—the beginning of a procedure, for instance—the design professional's representative can point out exactly what is acceptable and what is not. To do this after the fact or after the work is completed is not in the best interests of the project. It can very easily be seen that this can become an irritant and a problem to the operatives on the project.

Any time an interpretation or a decision or an expansion of information is required, it should be done promptly. If the contractors and workmen on the job know that the architect or engineer will be on the job at specific times, or within certain time frames, they can plan to approach him with some reliability and get the information they need. All information should, of course, be directed through the proper channels. The design professional should never talk to the subcontractors, vendors, or suppliers directly, but he should be talking to the manager, the contractor, or the people who have the proper lines of responsibility. Any communication, of course, should be backed up with written documentation and be properly distributed to keep all interested parties informed.

ON-SITE INSPECTION TECHNIQUES

The design professional's on-site representation can vary as widely as his clientele, their

```
1.03   QUALITY ASSURANCE:

       A.   Codes and Standards:

            Perform foundation drainage work in compliance with applicable
            requirements of governing authorities having jurisdiction.
```

Figure 10. Excerpt from typical specification. This represents an inadequate specification requirement. It is very vague and shows that the design professional is uncertain about what standards apply. This unfairly shifts the design burden to the contractor, and the professional loses control.

projects, and the sum of money available. Basically, the representation can involve the design professional in varying degrees, from none to constant on-site representation by a number of people. Some contracts that are awarded to design professionals do not include provisions for extensive on-site inspections. This is most unfortunate because an outsider unfamiliar with the intricacies of the design concept and its development must take the documents and attempt to learn the project so he can properly inspect it during construction.

The outsider, be he an individual, a team of individuals, or a separate company, loses the flavor of the project and can see only the technical mechanisms that are combined to form the project. In some instances this can be a tremendous drawback, not only to the project itself, but to the relationships that are established either by contract or through individual contract on the project. For example, an architectural firm designed a new school for a local board of education. The board chose to have the design professional prepare only contract documents and made no provision in the contract for on-site inspection. The architect involved happened by the project one day and, knowing some of the personnel from other projects, stopped by for a quick review and social visit. On entering the job trailer of the board of education, he was met by some old friends and some people he did not know. But the greeting was the same—a disgruntled, irritated one that bordered on unwelcome. The project had progressed to the point where the masonry was being installed, and the people involved were very bothered by the "lousy job" the architect had done in laying out and dimensioning the masonry for the building. The lines of dimension didn't add up properly, and the dimensions themselves were not modular to reflect the masonry involved. A great deal of extra work was being done by the inspectors to properly reflect the project requirements. After the inspectors had made their complaints known, the architect, thankfully a very concerned and

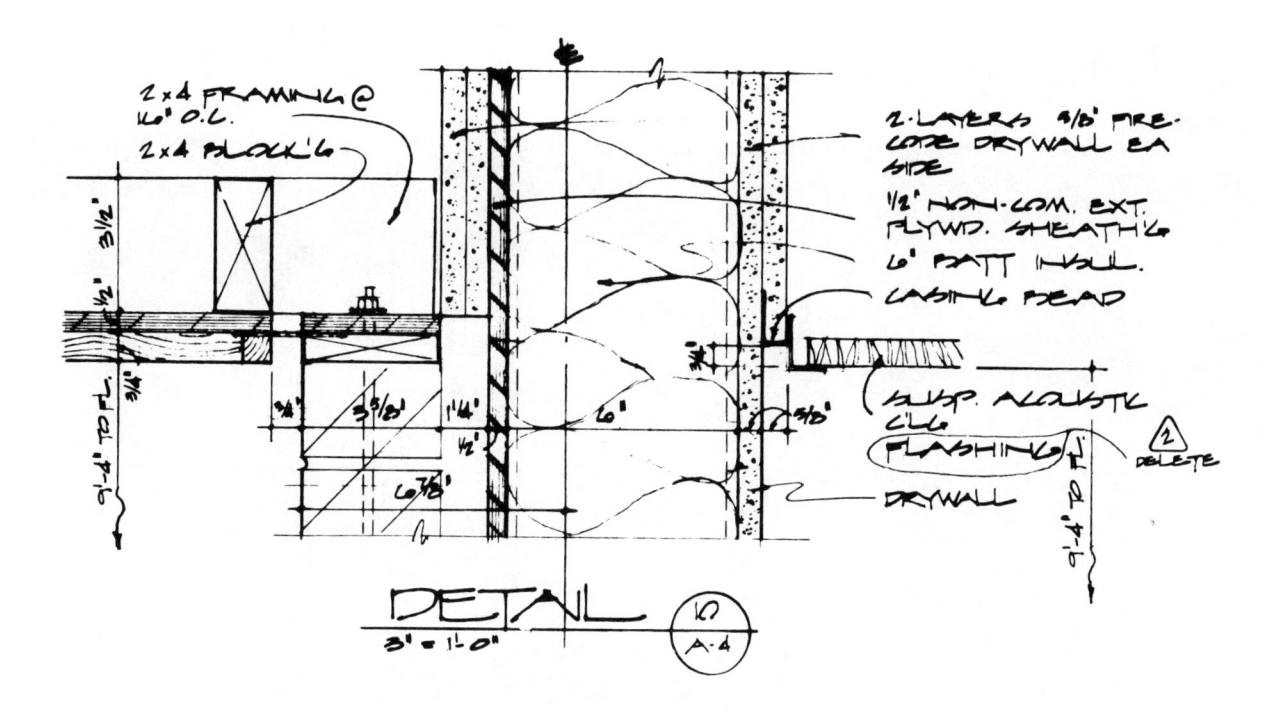

Figure 11. This architectural detail is for a limited area, but shows several items involved. This is a complicated piece of work for an inspector, several people will have to view this work when in place.

```
1.05 QUALITY ASSURANCE:
A.  STANDARDS:
    THE FOLLOWING STANDARDS ARE HEREBY MADE A PART OF THESE
    SPECIFICATIONS:

    1)  "BUILDING CODE REQUIREMENTS FOR RE-INFORCED CONCRETE",
        AMERICAN CONCRETE INSTITUTE (ACI 318-71).

    2)  "RECOMMENDED PRACTICE FOR MEASURING, MIXING AND PLAC-
        ING CONCRETE", AMERICAN CONCRETE INSTITUTE (ACI 614-59).

    3)  "ACI STANDARD RECOMMENDED PRACTICE FOR SELECTING PRO-
        PORTIONS FOR CONCRETE", AMERICAN CONCRETE INSTITUTE
        (ACI 613-54).

    4)  "RECOMMENDED PRACTICE FOR COLD WEATHER CONCRETING",
        AMERICAN CONCRETE INSTITUTE (ACI 306-66).

    5)  "RECOMMENDED PRACTICE FOR HOT WEATHER CONCRETING",
        AMERICAN CONCRETE INSTITUTE (ACI 605-59).

    6)  "MANUAL OF STANDARD PRACTICE FOR DETAILING RE-INFORCED
        CONCRETE STRUCTURES", AMERICAN CONCRETE INSTITUTE
        (ACI 315-65).

    7)  "RECOMMENDED PRACTICE FOR CONCRETE FORMWORK", AMERICAN
        CONCRETE INSTITUTE (ACI 347-68).

    8)  TESTING AS PER CURRENT ASTM STANDARDS.

    9)  BUILDING CODE OF THE CITY AND STATE IN WHICH THIS DEVELOP-
        MENT IS LOCATED.

B.  THE REQUIREMENTS OF "SPECIFICATIONS FOR STRUCTURAL CONCRETE
    FOR BUILDING" (ACI 301-72) ARE A PART OF THESE SPECIFICA-
    TIONS. THESE PROJECT SPECIFICATIONS DEFINE MATERIALS, ADDI-
    TIONS, CHANGES OR OMISSIONS FROM ACI 301-72 AND GOVERN OVER
    THEM. (ITEMS ARE NUMBERED ACCORDING TO RELATED CHAPTERS
    OF ACI 301-72.)
```

A. Codes and Standards:

1. Comply with the provisions of the following codes, specifications and standards, except where more stringent requirements are shown or specified:

    a. ACI 301 "Specifications for Structural Concrete for Buildings".
    b. ACI 311 "Recommended Practice for Concrete Inspection".
    c. ACI 318 "Building Code Requirements for Reinforced Concrete".
    d. ACI 347 "Recommended Practice for Concrete Formwork".
    e. ACI 304 "Recommended Practice for Measuring, Mixing, Transporting and Placing Concrete".
    f. Concrete Reinforcing Steel Institute, "Manual of Standard Practice".

Figure 12. Excerpt from typical specification. This lists the standards to which the concrete work must conform. Use of dates in the upper form may conflict with dates of the standards in the code. Lower form states standards and implies latest editions, which will usually agree with code provisions. Confusion can occur when various dated editions change requirements.

patient fellow, got into a deeper discussion with them. He was able, after a few minutes, to ascertain that the requirement that he had followed in the board of education's design manual to use standard modular brick as the basis for the masonry work had been changed. Indeed, the brick selected by the board was a norman-size brick, oversized and larger than the modular brick he had specified. Therefore, the problem was not his. He had, indeed, faithfully followed the directions of his client and produced a set of well-coordinated, properly dimensioned drawings.

Here it can be seen that the very basis for the misunderstanding could have made the architect very disgruntled. He could have stalked off, refusing all further contact with the board of education or at least could have spread tales about their operation. How easily one telephone call could have erased the consternation of all parties!

This story shows that on-site inspection by someone other than the design professional can be very hazardous. Today, when lawsuits are frequent and wide-ranging, including everyone involved in a situation, the design profes-

```
ABBREVIATIONS:

Reference to technical society's organization, bodies or documents
is made in specifications in accordance with the following
abbreviations:

ACI             American Concrete Institute
AGA             American Gas Association
AIA             American Institute of Architects
AIEE            American Institute of Electrical Engineers
AISC            American Institute of Steel Construction
ARI             Air-Conditioning & Refrigeration Institute
ASA             American Standards Association
ASHRAE          American Society of Heating, Refrigeration &
                    Air-Conditioning Engineers
ASLA            American Society of Landscape Architects
ASME            American Society of Mechanical Engineers
ASTM            American Society of Testing Materials
AWSC            American Welding Society Code
BOCA            Building Officials Conference of America
CSI             Construction Specifications Institute
FS              Federal Specifications
IES             Illuminating Engineers Society
NAAMM           National Association of Architectural Metal
                    Manufacturing
NBFU            National Board of Fire Underwriters
NBS             National Bureau of Standards
NEC             National Electric Code
OBC             Ohio Building Code
ODHS            Ohio Department of Highway Specifications
SHI             Steel Joist Institute
SPR             Simplified Practice Recommendation
UL              Underwriters Laboratories, Inc.
```

Figure 13. Excerpt from typical specification. This brief list of abbreviations and standards used on the project establishes the basis of the work and allows everyone involved to see what is required, but reduces the written material.

sional can be forced to pay a large sum of money simply to show that a problem that developed during construction was not his direct responsibility. He would have to show that his drawings and specifications were properly conceived and executed and that it was during the construction phase, over which he had no control, that the problem had developed. He would have to show that had his drawings and specifications been faithfully executed, the project would not have had this problem. But to do this, he would have to spend his own time and his own money.

Many design professionals will take a very hard-line approach to ensure that they have at least some on-site inspection capability. They want to be sure that their documents are executed properly. Numerous costly instances could be cited where the design professional was required to defend himself against litigation, because his documents were used, but he had no on-site, in-progress inspection responsibility. Unfortunately, lawsuits are drawn to include everyone involved, and some must pay dearly to provide a defense and an explanation of their responsibilities.

The traditional on-site inspection system revolves around the use of one person from the design professional's office who is familiar with the project or perhaps with several projects. Usually he is able, in touring around on a daily basis, to inspect and coordinate all the projects under his control. This system works very well for the small to moderate-sized office with projects of limited size.

When the office begins to grow and expand not only in the number of projects, but in the scope of projects as well, the one-person inspection system begins to prove inadequate. A very large project may demand so much of his time that he will be forced to neglect or give very cursory service to the other, smaller projects. Of course, this is a disservice to the clients.

When a firm is in this position, it must evaluate what steps it wants to take and what direction it wants to pursue; and it may very well find that the one-person inspection system becomes a team or even a separate department. With an entire department, inspection can become only a part of its responsibilities. It can involve all the shop drawing work, the issuance of the change orders, and other administrative matters. Help may have to be assigned to the inspectors to assist with this contract administration.

Although architects and engineers are not specifically or narrowly educated as inspectors, they find themselves "pigeon-holed" during their careers. It is very hard for someone who likes field work and enjoys the intricate detail of a job and the actual building of the project to, at some point, be drawn back into the office with collar and tie and forced into a design situation. He may be a fully registered architect or engineer, but his personality, experience, and the people he likes to associate with may very well be totally field oriented. This certainly is no particular stigma for the man involved, for if the field work (the actual construction) becomes shabby and shoddy and is not executed properly, crisply, and well, a very beautiful design concept can be totally destroyed.

It is important that every design professional's office have a well-trained, well-qualified, knowledgeable, flexible individual as its field superintendent or inspector. He must maintain deep regard for the design concept. At the same time, he must have the flexibility and the innate ability to understand the execution of the details of the drawings so that the work is done to enhance and support the basic design concept.

Many projects will require, or perhaps the owner will demand, that the architect have some sort of daily on-site representation. From biblical times, the person involved in such a situation has been known as the clerk of works. This is a professional hired by the architect or engineer, whose expenses and salary are paid by the owner in addition to the design professional's basic contract fee. The clerk of works will be on the site performing a wide range of duties, eight hours a day, every day. If there is any problem, the other operatives on the site know that the architect has a representative there at all times. To a large degree, he is able to solve the problems on site or with a consultation with the design professional's office. This is especially helpful when a project is very complex, if there are a number of operatives on the job, or where the construction is so massive and involved that only daily, ongoing inspection can possibly cover the work.

With the onset of construction management techniques, many architectural and engineering firms are offering their services as construction managers and literally have taken over the bidding, the coordination of bidders, the award of contracts, the scheduling of work, the expediting, and so on. In many firms construction management has become a separate department or even an auxiliary firm of the design professional's practice.

There is no particular conflict of interest; the design professional is working on his own project and attempting to provide the owner with extra services while still dealing with the same basic cadre of personnel. When the construction manager is a separate adjunct firm with its own staff, the management firm can service projects produced by other offices and can provide additional, low-risk (since a set fee is established) income to the parent firm.

The design professional must be fully aware that all his consultants must have representation on the job from time to time. Whether these consultants are dealing with landscaping, acoustics, or mechanical systems, there is a need for them to be represented. They must make their inspections, share their views, and give their input and feedback to the project and the various operatives. The design professional should see to it that all his consultants are contracted in such a way that these services are not provided at additional cost but are part of their basic services.

CHAPTER 3
THE CONSTRUCTION MANAGER

CONSTRUCTION MANAGER *n.* person, firm, or corporation who performs special management services during the construction phase of the project under separate or special agreement with the owner. May be an architect, engineer or others. This is not part of the architect's and engineer's basic services; it is an additional service sometimes included in comprehensive services.

GUIDANCE

In the past the vast majority of construction projects have involved a general contractor, a contractor for plumbing, one for heating, ventilating, and air conditioning, and one for electrical work. Traditionally, four separate contracts were set up between the owner and the contractors involved.

This setup has changed through the years, mainly because it has been seen that a lack of control on the construction site was detrimental to the project. Between the traditional configuration and construction management, now very popular, there were various other contractual situations. For example, the general contractor would have the mechanical contracts assigned to him for administration and coordination. For this extra work the owner would pay the general contractor a separate "administration" fee. Other projects had their contracts drawn so that there was only one contract, which involved all the various working trades.

The problem with the various configurations, however, was that many general contractors did not have the resources, expertise, or manpower to handle a large-scale administrative job. The general contractor, involved with all the architectural trades, had enough work on his hands controlling the various subcontractors, material suppliers, and others, pertaining to his own contract. Because of the work involved and despite the fact that he was paid an extra fee, the general contractor usually was not an effective administrator of the other contracts assigned to him. For this reason, the design professionals and the owners had to come to a different solution that would provide better results. At the same time, they did not want to increase the involvement of the professional's or the owner's staff in the construction project.

The system that was devised was given the name *construction management*. At its best, construction management is exactly that—a management and expediting function. Some of the first firms or individuals who moved into construction management were the general contractors. Basically, their experience in dealing with varying and numerous subcontracts stood them in good stead as construction managers. The element that changed, however, was that the general contractor gave up his on-site working function in that he himself did not have an active building function on the project. In essence, he was not a working contractor.

Taking away his participation in the actual construction and allowing him to function strictly as a manager greatly increased his effectiveness in administrating and managing the project. Most firms either became involved in construction management exclusively or expanded their operations to include a separate construction management function. Management is far less risky than general contracting, since the fee is set and guaranteed and no risk of massive financial loss is involved.

At the same time, many design professionals also moved into construction management so that they could achieve more control of the project itself. On a privately funded project the design professional-turned-manager could indeed control who bid on the project by accepting proposals only from highly qualified and exremely reliable contractors. By maintaining control over this actual selection process, and an even tighter direct control over the operatives on the job, the design professional felt that he could produce better projects—better in the sense that they met the budget with high-quality workmanship and materials within the time frame demanded by the owner.

This technique, however, must be refined and disciplined, so that tight control of the work does not bog down and literally strangle a project. Also, there is a danger of conflict of interest, which must be contained; impartiality is lost when the design professional is the construction manager. Generally, the management should come from a separate organization entirely.

The two management techniques are interesting in the contrast of their approaches. Although rooted in entirely different philosophies, both the contractor-manager and the design professional-manager can make the system work very effectively. It would seem that the contractor has the advantage in attracting and dealing with other contractors and subcontractors. The design professional seems to have the advantage in overall project control, including control of the owner.

But the final analysis shows that the construction management process in many instances will enhance the project by reducing the cost, controlling more tightly the quality of the project, and meeting much more accurately the time schedules.

It is extremely difficult for a superintendent employed by a general contractor to be in charge of overall supervision of the project, as well as trying to accomplish his own fair share of the work. Traditionally, a well-trained journeyman carpenter has been able to move into the superintendency of any construction project. Many general contractors, however, expect him to produce as much actual carpentry work as administrative work, but this is an impossible task. As construction projects were becoming more and more complex, there had to be a change; it is easily seen that construction management is the answer.

CONTRACTUAL OBLIGATIONS

The terms of the *General Conditions of the Contract for Construction* (in the American Institute of Architects construction management edition) are very specific when outlining the duties for each participant in the project, including the construction manager, but in the wording of the *General Conditions,* the manager comes by other duties by virtue of what is not written but is inferred in the *General Conditions.*

The *General Conditions* call specifically for the construction manager to be hired to manage the construction of the project. His prime obligation, by virtue of these contract documents, is the actual expediting, scheduling, and coordination of all the work and contractors working on the project. As has been discussed, the construction manager is generally replacing the traditional general contractor.

In his position of scheduling and coordinating the work, the manager, along with the architects, is ultimately responsible for reviewing and processing the certificates for payment, including final payment. The manager actually makes recommendations to the architect who issues the proper certification. Here is one of the first subtleties in the *General Condition* language. Certainly no construction manager is going to make any sort of recommendation, with regard to payment, if he does not have some substantiating information for that recommendation. Basically, the contract language implies that he must inspect the work. In his recommendation he verifies his observations with the architect who can issue the certification. This means that the manager must tell the architect that certain work has been accomplished and is satisfactory and that payment

24 Systematic Construction Inspection

ARTICLE 2   ARCHITECT AND CONSTRUCTION MANAGER

2.1   Definition

2.1.1   The Architect is
                          and identified as such in the Agreement and is
             referred to throughout the Contract Documents as if singular in number
             and masculine in gender. The term Architect means the Architect or
             his authorized representative.

2.1.2   Nothing contained in the Contract Documents shall create any contractual
             relationship between the Architect and the Contractor.

2.1.3   The Construction Manager is
                                     who will provide construction management ser-
             vices as hereinafter described or referred to in the Contract Documents.
             The term "Construction Manager" means the Construction Manager or his
             authorized representative.

2.1.4   The Construction Manager will be responsible for Scheduling and coor-
             dinating the work of all Contractors on the Project.

2.2   Administration of the Contract

2.2.1   The Architect and Construction Manager will provide general Administration
             of the Construction Contract, including performance of the functions
             hereinafter described.

Figure 14. Excerpt from typical specification. This insert delineates the duties of the architect and the construction manager. It is helpful in showing all participants the areas of responsibility and the lines of communications.

should be made to the contractor responsible. Although there is not a specific mention of the word "inspection," the responsibility to inspect is still implied within this particular portion of the *General Conditions.*

Further, the *General Conditions* specifically refer to the work of the architect and his responsibility with regard to on-site observations. They call for noncontinuous on-site inspection. The construction manager, however, is on the site daily, much as the clerk of works is. Through his recommendation for payment certification and his obligations to reject unsatisfactory work, he must be part of the inspection team. The *General Conditions* specifically mention that the construction manager may reject work that does not conform to the contract documents. If his inspection reveals faulty work, a recommendation listing the work and various preferable solutions is prepared, including any impact on the job as a whole. His recommendation, again, is to the architect, who will inspect the same work and issue any particular orders to the contractors. The construction manager is, therefore, a constant on-site inspector, although his enforcement authority is somewhat restricted.

This is the crux of the construction manager system: constantly keeping his eyes on the job, so that he can watch each item of work as it proceeds and advise the architect as to the amount and satisfaction of the work involved. Construction managers may also require special inspection or testing of any work that he feels does not comply with the contract documents. He must, of course, have reasonable cause for such action, and it is subject to the interpretations and decisions of the architect. Again, this is a situation where the architect's contract calls for a limited level of inspection. By virtue

2.2.2   The Architect and Construction Manager will be the Owner's representative during construction and until final payment. The Architect and the Construction Manager will have authority to act on behalf of the owner to the extent provided in the Contract Documents, unless otherwise modified by written instrument which will be shown to the Contractor. The Architect and the Construction Manager will advise and consult with the Owner and all of the Owner's instructions to the Contractor shall be issued through the Construction Manager.

2.2.3   The Architect and the Construction Manager shall at all times have access to the Work whenever it is in preparation and progress.

The Contractor shall provide facilities for such access so the Architect and the Construction Manager may perform their functions under the Contract Documents.

2.2.4   The Architect and the Construction Manager will provide a full-time representative to generally observe the progress and quality of the Work and to determine in general if the Work is proceeding in accordance with the Contract Documents. On the basis of their on-site observations as Architect and Construction Manager, they will keep the Owner informed of the progress of the Work, and will endeavor to guard the Owner against defects and deficiencies in the Work of the Contractor. The Architect and the Construction Mangger will not be responsible for construction means, methods, techniques, sequences or procedures, or for safety precautions and programs in connection with the Work, and they will not be responsible for the Contractor's failure to carry out the Work in accordance with the Contract Documents.

2.2.5   Based on such observations and the Contractor's Applications for Payment and recommendations of the Construction Manager, the Architect will determine the amounts owing to the Contractor and will approve Application for Payment in such amounts, as provided in Paragraph 9.4.

2.2.6   The Architect will be, in the first instance, the interpreter of the requirements of the Contract Documents and the judge of the performance thereunder by both the Owner and Contractor. The Construction Manager will, within a reasonable time, render such interpretations as he may deem necessary for the proper execution or progress of the Work.

2.2.7   Claims, disputes and other matters in question between the Contractor and the Owner, relating to the execution or progress of the Work or the interpretation of the Contract Documents shall be referred initially to the Architect through the Construction Manager for decision which the Architect will render in writing within a reasonable time.

2.2.8   All interpretations and decisions of the Architect shall be consistent with the intent of the Contract Documents. In his capacity as interpreter and judge, the Architect will exercise his best efforts to insure faithful performance by both the Owner and the Contractor and will not show partiality to either.

2.2.9   The Architect's decisions in matters relating to artistic effect will be final if consistent with the intent of the Contract Documents.

Figure 14. (*Continued*)

2.2.10   Any claim, dispute or other matter that has been referred to the Architect, except those relating to artistic effect as provided in Subparagraph 2.2.9 and except any which have been waived by the making or acceptance of final payment as provided in Subparagraphs 9.7.5 and 9.7.6 shall be subject to arbitration upon the written demand of either party. However, no demand for arbitration of any such claim, dispute or other matter may be made until the earlier of:

2.2.10.1   The date on which the Architect has rendered his written decision, or

2.2.10.2   the tenth day after the parties have presented their evidence to the Architect or have been given a reasonable opportunity to do so, if the Architect has not rendered his written decision by that date.

2.2.11   If a decision of the Architect is made in writing and states that it is final but subject to appeal, no demand for arbitration of a claim, dispute or other matter covered by such decision may be made later than thirty days after the date on which the party making the demand received the decision. The failure to demand arbitration within said thirty days' period will result in the Architect's decision becoming final and binding upon the Owner and the Contractor. If the Architect renders a decision after arbitration proceedings have been initiated, such decision may be entered as evidence but will not supersede any arbitration proceedings unless the decision is acceptable to the parties concerned.

2.2.12   The Architect and the Construction Manager will have authority to reject Work which does not conform to the Contract Documents. Whenever, in his reasonable opinion, he considers it necessary or advisable to insure the proper implementation of the intent of the Contract Documents, he will have authority to require special inspection or testing of the Work in accordance with Subparagraph 7.8.2 whether or not such Work be then fabricated, installed or completed. However, neither the Architect's authority to act under this Subparagraph 2.2.12, nor any decision made by him in good faith either to exercise or not to exercise such authority, shall give rise to any duty or responsibility of the Architect to the Contractor, any Subcontractor, any of their agents or employees, or any other person performing any of the Work.

2.2.13   The Architect will review Shop Drawings and Samples as provided in Subparagraphs 4.13.1 through 4.13.8 inclusive.

2.2.14   The Architect will prepare Change Orders in accordance with Article 12, and will have authority to order minor changes in the Work as provided in Subparagraph 12.3.1.

2.2.15   The Architect and the Construction Manager will conduct inspections to determine the dates of Substantial Completion and final completion, will receive and review written guarantees and related documents required by the Contract and assembled by the Contractor and will issue a final Certificate for Payment.

Figure 14. (*Continued*)

> 2.2.16 The Architect will provide one or more Full-Time Project Representatives to assist the Architect in carrying out his responsibilities at the site.
>
> The duties, responsibilities and limitations of authority of such Project Representative shall be as set forth in AIA Document B352.
>
> 2.2.17 The duties, responsibilities and limitations of authority of the Architect and the Construction Manager as the Owner's representative during construction as set forth in Articles 1 through 14 inclusive of these General Conditions will not be modified or extended without written consent of the Owner, the Contractor and the Architect.
>
> 2.2.18 Neither Architect nor the Construction Manager will be responsible for the acts or omissions of the Contractor, any Subcontractors, or any of their agents or employees, or any other persons performing any of the Work.
>
> 2.2.19 In case of the termination of the employment of the Architect or the Construction Manager, the Owner shall appoint an architect or a Construction Manager against whom the Contractor makes no reasonable objection, whose status under the Contract Documents shall be that of the former architect. Any dispute in connection with such appointment shall be subject to arbitration.

Figure 14. (*Continued*)

of the responsibilities given to the construction manager, he is in the position of being an inspector although not specifically called that in the documents.

The contract between the construction manager and the owner is parallel to that between the architect and the owner. There is no contractual obligation of any sort between the architect and the construction manager. Although both are basically agents of the owner, they have specific responsibilities to the owner, and these responsibilities interlace in only a few places. The interlacing is primarily there for overall quality control or inspection and for the expediting and proper execution of the project. This last phrase is, perhaps, the key to the entire system and the project. The contractual lines are devised so that the project is of primary importance, in full keeping with the owner's concern. The details of how his project is executed, or who executes it, really are of little interest to the owner. He is interested in the end product, in occupying the project on time, and in not exceeding the budget. The owner has tremendous and wide-ranging interests beyond the construction itself. He is concerned about moving his operation, furnishing the new facility, renting available areas in the building, maintaining business while moving, and so forth.

The architect, with his many interests on a particular project, simply does not have the time to assume the role of construction manager. Although many architectural firms may have a construction manager function, this usually becomes a separate department or organization and a separate contract. Besides removing conflict of interest, this allows the proper allotment of time for each function. In the past, the architect in his classic role of master builder, often tried to control a project by acting as the construction manager before this nomenclature became popular. If he is in a position of having no other work in his office and no other clients to whom he has obligations, this situation may work. Because of his role in any one project, and because of the many other projects in which he may be engaged, he simply does not have time to become the one who expedites and schedules each project. He is, however, interested in how this is done. But to have someone else, with another set of eyes and another set of hands, to set all this in motion is a valuable aid for the architect.

The worst possible situation would occur on a project that has an architect and a construction manager who simply do not get along and cannot work together. This is quite possible, but it is the owner's responsibility to prevent it from happening. In using the architect as his basic advisor, the owner can develop a team (which includes selection of the manager) that can work together and keep the project as the prime goal. There is so much potential for controversy within any project that it is almost mandatory that the manager and the architect be the most compatible of participants on the job. This is not to say that they work with such a hand-to-glove relationship that their respective responsibilities and functions become cloudy. They have to work together, perhaps daily, in seeing that the work is done properly. There is no reason the construction manager and the architect cannot talk to the owner in different ways; indeed, at times they may be in direct opposition to each other. Many different approaches can be taken to best serve the owner and his project, and therefore, communication with the owner, like all the communication on the job, should be open and active.

It is highly improper for the manager or the architect to go to the owner and criticize the other, in such a manner that the owner becomes suspicious of both participants. Before long, there can be pique on either part, which may become openly hostile and cause controversy on the entire project.

The owner must understand the wording and the lines of responsibility laid out in the *General Conditions*. It may even be proper for the architect to take an extract from the *General Conditions* and list the individual responsibilities of the manager and the architect for the owner. In this way, the owner knows exactly whom to approach when he has a particular problem. It also helps to clear up some rather vague language of the *General Conditions*. Although the manager has no specific inspection responsibility under the *General Conditions,* except in dealing with substantial completion, they imply that he has an inspection function. He may not have the authority to deal directly with the problems he uncovers but in producing the recommendations to the architect for payment, remedial work, or additional testing, he must know what he is talking about and where the fault lies. To do this, he must become an inspector at least during part of his day.

Obviously, the construction manager system does not work, or is not necessary for every project, especially moderate to small projects. Larger, more extensive projects benefit greatly by using it. Without the manager's constant on-site presence, certain responsibilities of the architect are enlarged. He, of course, must be able to verify the work and properly certify it for payment. Therefore, the architect must assume some of the construction manager's subtle inspection obligations.

AVOIDING IMPEDIMENTS OF WORK PROGRESS

Because the construction manager's system is more complicated, it becomes eminently more important that all participants understand their specific roles. What are their responsibilities? What do their responsibilities exclude? The concept of "the project" must be kept in mind at all times. Any system that is imposed on a project should go to great lengths to avoid impeding the progress of work. This is particularly true of the construction manager system.

There is no doubt that the more participants there are in the construction and inspection system, the greater the chance for conflict, misunderstanding, and other impediments. Because of the contractual obligations of the manager and the architect and the parallelism of those contracts, it is essential that there be an extra effort to provide a smooth working relationship between the two. Communication should be comprehensive, constant, daily, very open, and complete.

If the manager, in doing his job, inspects some work and finds it faulty, he has no real recourse. His reaction must be to make a recommendation to the architect, who must "pick up" the situation and resolve it. If there is to be an interpretation of the contract documents, that function belongs by contract, solely to the architect. The manager must take responsibility if there is some fault in the scheduling or the management of the work. It is not enough to understand the particular roles; one must be able to understand how the process is imposed, how the responsibilities are interwined, and where decisions must be made. There are enough problems with delivery and shortage of materials, labor union contracts, the intolerance of people, and the like, on any construction job that the coordination between the manager and the architect cannot be allowed to deteriorate in any way. They must work together very closely in a complementary way, and they must understand and respect the process in which they are participating.

The architect, of course, has his inspection facility greatly enhanced by the construction manager system. The manager, if one were to use street language, could be called a "stool pigeon" for the architect, as he becomes an on-site inspector who can help the architect. In the end, he is of great help to the project itself.

Previously, the clerk of works concept was discussed, and this role, of course, can be added to the construction manager system. A very large project could very easily have an architect, his clerk of works, and a construction manager. This system is more complicated and deserves much more attention by the participants. New lines of communication must be established and maintained. The owner talks through his architect to the manager, who talks to the contractors. (The owner does not talk directly to the contractors.) The architect cannot give specific instructions to the contractor through his clerk of works without informing the construction manager, otherwise the system will begin to break down.

This introduction of more personnel into the inspection system cannot be allowed to stand in the way of the construction process. A delicate balance should be achieved so the project is not administration bound. Too many people acting in similar capacities tends to cloud solutions and create communication problems, and attitudes can change because someone is bypassed or his functions superseded.

There must be quality control and inspections. For the most part, the contractors respect and understand this process as it involves them. The contractor becomes concerned only if that system impedes his work and causes time delays because decisions are not made promptly or by the proper people; in such cases, decisions must sometimes be countermanded, reissued, or delayed still further by other sorts of entanglements.

With more complicated construction projects, the construction manager system works, so long as all participants understand the manager's emerging role—primarily, his contractual role in the process.

CHAPTER 4
THE OWNER

OWNER 1. The design professional's client. 2. The owner of a project, such as a homeowner or a governmental agency. 3. The person, firm, or corporation with which a contract has been made for the payment of the work performed under that contract.

PROTECTION OF INTERESTS

The owner involved with any particular construction project is vitally interested in the building process but, at the same time, he may be ignorant of what is happening. His interest, of course, follows on his expenditure of a tremendous amount of money. Whether the owner is an individual, a corporation, or a governmental agency, none wants to pay for a project not worth the money expended.

The owner for the most part wants dearly to be involved in the construction process, but it is the wise owner who sees how inadequate he is to the task. Except in rare instances, he would do well to involve a design professional whom he can work with, talk to, and confide in. His design professional, in serving him properly, becomes his agent, in most instances acting for him on the project and, because of contractual obligations, trying to protect his interests.

The owner, however, may very well see the design professional, even if he has full confidence in him, as just another person in the process who may try to invade the owner's interest. An error of any sort by a design professional may drastically change the project for the owner. A decision that the design professional makes may not be 100% in keeping with what the owner intended or anticipated. So, above all, the design professional has the responsibility to gain as much confidence from the owner as possible and to carry out his entire contractual agreement, to the letter, in a very open and complete manner.

The owner, however, still may want to be intimately involved in the project. If he is a homeowner, for his own satisfaction he often may visit the site nightly after the workers have gone, checking on progress and quality. He is really, in a sense, finding out what he is paying for. An owner who walks into an existing building and buys it as it stands does not have the problem of seeing the project in the "rough" and following the actual finishing process. He buys it as a package, much as he would buy a suit of clothes off the rack. Flaws may exist, but they are not nearly as apparent as when they are exposed one at a time during the fabrication. So it is with construction projects.

Almost everyone has some knowledge of construction and how things are put together. The owner may have helped build his childhood treehouse, but probably has not seen other buildings being built. In his adult working years, the owner has certainly seen things in buildings, some of which work for his business and some of which do not. It is to the architects' benefit if the owner has experienced a wide range of solutions or at least has some flexibility. The owner, at any rate, forms certain ideas about the best way to accomplish his project.

It makes no difference whether this is a fairly modest home, a corporate headquarters, or a large governmental complex; owners have a tendency to draw certain conclusions.

The agents of the owner (staff members, not the contractual agents such as the architect) may also be on the job. They too may have preconceived ideas of what the project should be. In many instances, the owner's little knowledge can, according to the traditional axiom, be dangerous. The design professional must be able to take the knowledge of the owner, control it, and channel it properly. This is done by preventing confrontations between the owner and the contractor, by processing complaints, questions, and inquiries, or by educating the owner the architect literally taking the owner by the hand and explaining the process.

In many instances, the owner can become frustrated to the point that he feels he is strictly the money man. He is giving up a tremendous sum for a commodity that he doesn't know much about, and he really isn't sure what he is getting. The homeowner, for the most part, unless he wants a custom-built home with a design professional involved, has very little in the way of help to call upon. He must rely on himself and the builder or perhaps on a friend or relative who knows a little something about building. The reliable builder will take on some of the responsibility of the design professional and will helpfully educate the owner whenever an inquiry is placed. The unscrupulous homebuilder will, of course, try to gloss over, ignore, or cover up work that is shoddy or substandard. Even if the owner has his "inspector's hat" on, he may be talked into accepting some work that is not in his best interest.

On large projects, the corporate and governmental owner may assign some of their staff members to oversee the construction. This, of course, is direct control by the owner. The representatives can attend job meetings, be consulted on any major decisions that are made, and be a member of the actual control team of the project. There is really no inherent problem with this type of control. The owner, however, and his representatives must understand, as would any other participant, the system in which they are participating. They must also be willing to give up any direct communication with the contractor, communicating only through the proper channels of the control team.

It is true that the owner has a direct contractual agreement with the contractor, and it would seem that this direct line of contract obligation would allow the owner to contact the contractor directly. However, this could lead to confusion on the project and could greatly impede the progress of the work. Direct contact would completely disrupt the communication system and would leave the design professional or the construction manager in very awkward positions; they would not know what the owner told the contractor, and it is very difficult to evaluate or control something you know nothing about. An aggressive owner must be controlled. He must be told emphatically in no uncertain terms, that he must work through his design professional and the established lines of communication.

A project team that has lost the owner's confidence (or never gained it in the first place) is in deep trouble. If the owner attempts to circumvent it, the entire team will be rendered ineffective, resulting in a very expensive nonentity (and a solid basis for litigation). An owner who chooses to deal directly with the contractor works to his own detriment in the long run. He should be made to see the tremendous advantage of coordinating his efforts and communications with and through the project team, starting with the design professional. It cannot be emphasized enough how important it is for the control team to have the owner's confidence and to have the owner understand that he must work through that team and not go to the contractor directly.

POLICING THE CONTRACT

The owner may be party to a number of contracts on any given project. There may be one with the design professional, one with the construction manager, and one, of course, with the contractor or on some projects, contractors. His primary effort on the project is one of paying off these contracts as the terms are met. The owner is the one common denominator of all these contracts, being party to all of them. There are no contracts between the architect and the manager, the manager and the contractors, the contractors and the architect, and so forth.

The owner can be likened to a driver of a large team of horses, holding many reins in his hand, trying to control each horse so that they pull together doing the work required by the driver. This analogy, to the owner means that he must work with the architect, the manager, the contractors, and the other contractees individually, while at the same time pulling them together into a working team to complete his project.

Any misunderstanding, harrassment, or adversary action between any participant and the owner can only be an irritant to everyone involved and can greatly impede the work. For example, an owner who becomes disenchanted with the consulting engineer on his project will have an extremely hard time working with this individual. However, he would also put his other design professionals in a uncomfortable position. Although the owner does not have a direct contract with his consulting engineer, the architect does, and the architect is then in the peculiar position of trying to maintain his credibility with the owner, keeping his engineer under control, and at the same time maintaining the progress of the project.

This is not to say that the participation of the owner is unwanted or unnecessary. But when the owner chooses to participate, he is well advised to understand the project as a whole and to understand what his comments and his desires might produce if they are misdirected.

ARTICLE 3   OWNER

3.1   **Definition**

3.1.1   The Owner is the BOARD OF DIRECTORS OF
identified as such in the Agreement and is referred to throughout the Contract Documents as if singular in number and masculine in gender. The term Owner means the Owner or his authorized representative.

3.2   **Information and Services Required of the Owner**

3.2.1   The Owner shall furnish all surveys describing the physical characteristics, legal limits and utility locations for the site of the Project.

3.2.2   The Owner shall secure and pay for easements for permanent structures or permanent changes in existing facilities.

3.2.3   Information or services under the Owner's control shall be furnished by the Owner with reasonable promptness to avoid delay in the orderly progress of the Work.

3.2.4   The Owner shall issue all instructions to the Contractor through the Construction Manager.

3.2.5   The foregoing are in addition to other duties and responsibilities of the Owner enumerated herein and especially those in respect to Payment and Insurance in Articles 9 and 11 respectively.

Figure 15.   Excerpt from typical specification. This specification insert shows the areas where the owner will be active and responsible.

There is no way that an owner can be kept away from the job or out of the process. If the participation of the owner becomes a cumbersome, time-consuming impediment to the work, then he is working against himself and he has no one to blame for budget overruns and the missing of time deadlines. Of course, the owner will not always see this and will try to blame other participants. We can only advocate that the owner think carefully about just how he wants to participate. That participation should be based on his confidence in the other contract members.

Corporate and governmental owners may very well have a policy or statutory requirement to participate in and oversee the project. But again, there is always a supervisor over an owner's representatives. The supervisor and the representatives should all understand exactly what they can do, what they can gain and lose, and how their participation can directly affect the work.

DECISION MAKING

In instances where the owner is represented on the project, the major benefactor of that presence is the decision-making process. With constant representation on the project and with authority to act for the owner directly, the representative can greatly expedite and simplify decision making.

The representative of the owner must, of course, be thoroughly familiar with the inspection process and must actively participate in it. The representative simply must keep abreast of the progress of the job. And it is this representative who has the luxury to reinspect the work at his leisure to be sure that there has been no deterioration and that work previously installed is, indeed, still installed properly.

Construction sites are often vulnerable to a great deal of vandalism and damage. It is good that the owner's representative is able to inspect the job over and over again, including work that has been previously completed. This is not to say that this work is totally ignored by others on the inspection team. But basically inspection is ongoing, and the other participants do not have the time for reinspections. With his proper inspection process, the owner's representative can, indeed, keep abreast of the job and may, for the most part, be ahead of the job if he can foresee what is to come and what problems are at hand.

The owner's representative should not replace any member of the inspection team, but he is an additional member with another point of view and a different interest, serving the decision-making process as well. The delays of communicating with the owner are minimized, and a representative, if he is properly authorized, can make ongoing, immediate decisions as the job progresses, saving time, money, irritation, and involvement to all on the project. An owner who may be far away from the project can present a problem to the progress of the job. If decisions are not forthcoming immediately to meet the conditions of the job, contractors will often put in claims for extra time based on the interval needed to receive the decision.

This area can be a source of irritation, aggravation, and, indeed, even hard feelings, because it is never the owner's intention to purposely delay the job. Even simple communication processes, such as telephone, telegraph, or mail, can break down, thus impeding the job. Sometimes, too, an owner remote from the project will not be thoroughly abreast of the job progress and will need time to become informed. He may have to study the drawings, previous reports, and so forth, before being able to make a good and proper decision.

Delays can occur again and again, leading to an added cost of time, personnel involvement, rentals, interest on loaned money, and so forth, to complete the job. Today, when money is extremely tight and profit margins are very small, the decision-making process must be expedited where at all possible. Perhaps one of the earliest ways is with the on-site representation of the owner.

COST CONTROL

The importance of cost control in the modern-day construction process cannot be emphasized enough. Numerous projects have run into a good deal of trouble because of the cash flow problems of the contractor, delays in decision making, a general slowdown of work, and the inspection process. The owner's representative can, of course, be one of the major forces behind an aggressive approach to the proper progress of a job. His major interests are to prevent extra charges of any kind to the owner and to deliver the job properly and on time. Keeping these two factors uppermost, coupled with the decison-making process, control of the costs can be achieved, but it requires all members of the inspection team to fully participate and to work closely with one another. It is essential that mandatory inspections be made promptly and properly, which in turn means that the work to be inspected must be executed, finished, and made ready for inspection according to schedule.

All the participants in the inspection process have some financial interest in the job itself, either for themselves or for their agencies. A building inspector, for instance, coming onto the job only to find that no progress has been made since his last visit is wasting the taxpayers' money. Job conditions will dictate changes in progress, but the building inspector should have enough knowledge of the job so that he schedules timely inspection trips. The owner is interested in seeing that things are inspected so that the progress of the job is not impeded or stopped because of faulty work, and so on for each participant.

Everyone endorses progress, but inspection must accommodate it. It is to no one's benefit to have inspection lag. Eventually some faulty condition will be found, which may then require work stoppage and remedial action. If inspection lags and allows work to move substantially beyond the faulty condition, repair work and cost factors can be greatly effected. In such cases the inspection system has totally broken down in relation to the total project.

Cost control in modern construction can be one of the most confounding factors. Many are quick to allow the design concept to dissipate for the sake of the job progress and saving money as the project proceeds. Although this practice should be minimized or not permitted at all, it is constantly at work on the construction site. Whether the project is being delivered as originally conceived depends on the dedication and commitment of the owner and his representative. Many times the representative will allow or actually force work to be changed for cost control purposes. These changes, if made without full coordination with all members of the construction and inspection team, are a tremendous source of confusion and irritation. This is particularly true of the design professional and the contractors involved. There is no indication that the project is short-changed at this juncture. It must be pointed out, however, that for cost control to be truly effective it must be an intimate part of the overall project control and decision-making processes. That, in turn, is an intimate part of the inspection system.

CHAPTER 5
THE CONTRACTOR

CONTRACTOR *n.* 1. The individual, firm, or corporation undertaking the execution of the work under the terms of the contract and acting directly or through its agents or employees. 2. A person or company who agrees to furnish materials and labor to do work for a certain price.

CONTRACTUAL OBLIGATIONS—THE MANY THINGS INCLUDED

The contractor, by virtue of the contract that he signs, has a tremendous amount of responsibility for the execution of the project. The architect, while designing the project, has gone through a myriad of processes in which he has selected various materials, components, and systems for incorporation into the project. He has made an overwhelming number of decisions, large and small, about how the project should be constructed and what should be included. It is said that perhaps a half million different pieces of construction apparatus are involved in even a moderate-sized project.

Although there is a tremendous amount of responsibility during the design phase, the responsibility carried by the contractor begins with the architect's decisions. The drawings and specifications, which depict all the items, devices, apparatus, components, equipment, and materials, are basically a shopping list of what the contractor must purchase, install, or at the very least, coordinate. An astute contractor becomes well versed in and thoroughly familiar with the contract documents. He may be more familiar with what is not included in the contract documents. These loopholes can provide him with a source of additional income because of change orders, addenda, and other documents that must be issued to either clarify or make additions to the project.

The unscrupulous contractor may try to take advantage of this situation by charging inflated prices for the additional work. He knows full well that the work is needed and will react to this unusual supply and demand situation. For the most part, though, the contractors will cooperate and provide the additional work in the normal sequence without extensive negotiations and exaggerated prices. They know it is for the benefit of the project, and in most instances they will call errors to the attention of the project team.

The contractor cannot in any way neglect what is included because those things become his responsibility. If something is required of him and does not appear in the work, it is the contractor's responsibility to correct that situation. On a large project it is an awesome task for the contractor's staff to keep track of all the material and personnel and the proper scheduling of the job.

The contractor has a great many involvements in any project. He must organize his staff to follow the project from its very beginning to its final completion and occupancy. He must organize his staff in such a way that

36 Systematic Construction Inspection

> CODES AND STANDARDS:
>
> Where codes and standards are referred to, they shall be current approved copies. It shall be the duty of the supplier of any material on this work to submit evidence, if required, that his material is in compliance with the applicable local and state codes and standards, in the method in which the material is used in this project shall be current approved copies.
>
> Contractor shall submit his base bid in accordance with plans and specifications. If plans and specifications do not comply with any codes or utility company requirements having jurisdiction, then Contractor shall submit an alternate price on any changes necessary to comply with such codes. If such alternates are not stated in bid, it shall be assumed that Contractor's base bid includes all work necessary to comply with such codes or utility company regulations, and no extra shall be paid for any work or materials in order to meet requirements of codes or utility company regulations having jurisdiction.
>
> PERMITS AND FEES:
>
> Before any work is started or materials purchased or purchase commitments made, Contractor shall first take out and pay for all permits, licenses and fees as specifically detailed and required under the Building or Zoning Ordinances of the Municipality or the legal authority having jurisdiction. Copies of all such permits, etc., shall be submitted to Owner with first Application for Payment. EXCEPTION: State approval will be obtained by the Architect, with Contractor reimbursing the Architect for the State permit fee.
>
> In the event the necessary permits referred to herein cannot be obtained by Contractor, the Owner may without prejudice, terminate the employment of Contractor and declare contract of no further force and affect. Such action shall relieve Owner and Contractor of all obligations under contract.

Figure 16. Excerpt from typical specification. This insert firmly sets a basis for the work and the responsibility of the contractor showing pointed and detailed requirements.

some people will be working on up-to-the-minute projects, some will be working months ahead, and some phasing out projects months after completion. In this way the contractor can anticipate situations that are coming to the fore, as well as properly executing the work as it progresses. At the same time he is following up on details and finalizing those items that have been executed and passed on. Without this overall organization of the project, the contractor can become inundated and lost, and the project would suffer.

It can easily be seen that one major function of the contractor's staff is inspection. It is important that they be in a position to schedule, expedite, critique, order, pay for, store, move, and cajole as necessary. All these functions come down to the bottom line—inspection. If the contractor has not inspected the project continuously, he will not know how the project has progressed, where it is going, and where it has been, and he will lose control and be unable to execute the project to meet the terms of the contract documents.

1.05 COORDINATION OF THE WORK:

A. The contractor for general construction shall coordinate all work in progress on the site. He shall direct arrangements for the storage of materials; shall keep himself informed of the progress and detailed work of all contractors and shall work with the Architect in the coordination and expediting of all work so that the progress of the work shall be kept on schedule.

B. Contractors for all other items of the work shall cooperate with the contractor for general construction and with each other and shall keep informed of the progress of ALL the work and shall expedite and coordinate their work so that the progress of work AS A WHOLE is kept on schedule.

C. Each contractor shall confer and cooperate with all other contractors whose work occurs in the same area. If this contractor installs any of this work without such cooperation, and in doing so interferes with or prevents the installation of other work in the area, he shall bear

Figure 17. Excerpt from typical specification. This provision delineates how the work is to be coordinated and who is to act as coordinator, but the design professional is still total project coordinator.

The contractor may have several teams of people who are executing various forms of inspection. He may be involved with various subcontractors or even with the design professionals in a joint effort of inspection to ensure that the project is able to progress and will not become stalled because of insufficient work space or material or because previously executed work has proved faulty.

Contractors, for the most part, decry inspection. Many say it is unnecessary; many say it is an impediment; many say it is second guessing. Ironically, contractors are important participants in the inspection process. If their other cries of woe are indeed well founded, then the inspection process has broken down.

There is a tremendous need for contractors to meet with building officials on a continuing basis. Both have problems that involve the other, and discussion will solve many if not all of them. Semiannual seminars would be an excellent vehicle for discussion, as well as contractor membership and participation in building officials' groups. The trades should also realize that the building officials, in many instances, do not write the codes. Model codes, used widely in this country, are open to any challenge and the code writing groups are receptive to code change proposals from any person or group. We must get away from flexing "ego muscles" on both sides. We should be developing a deep concern and a dedicated effort in the design/build teams. Quibbling, hassling, and budgetary cheating over code provisions is a macabre exercise; it's gambling lives against money. Overall regulation of the construction industry is prohibitively expensive, but for the most part building code provisions are not the factors that add substantial cost to the building. In fact, the added cost is small when compared to the liability involved. Code items are not added work, they are basic work.

The contract makes the contractor walk a very narrow line. He is receiving a tremendous sum of money for the project, but little of it can be turned into profit. He and his staff are to order the material and to have it on the site in good condition, ready to be installed. He must have enough manpower and subcontractors to execute the work properly. Failure in any of these areas can reduce the project to chaos.

In addition, the astute contractor will quickly realize that he has some element of the inspec-

38 Systematic Construction Inspection

> COORDINATION OF TRADES
>
> a) Sections of these specifications set down guidelines as to the extent of work by subcontractors. These guidelines are set forth to aid in the bidding process and help with estimating trade payment breakdowns and monthly payment requests only.
>
> b) The General Contractor is responsible for the administration of all subcontractors and therefore is not obligated to the work descriptions listed herein. As previously mentioned the General Contractor must provide a total and complete project and thus may divide the work between subtrades as he feels is most efficient and economical to provide a complete project.
>
> c) The Architect is not responsible for providing complete coverage of work between subtrades in the work descriptions of the sections of this specifications. Therefore, the Architect will not arbitrate or even discuss overlapping or voids that occur in scope of work between subtrades. The General Contractor shall solely handle all problems of this type that deal with which subtrade will take care of which items of work.
>
> d) The general contractor shall coordinate all subtrades regardless of the specification headings and shall make all necessary provisions for accommodation of all equipment and fittings into the building and patching after necessary. The general contractor shall coordinate all material delivery, unloading and storage. These provisions do not however relieve the subcontractors from the responsibility of coordinating all work with the other subtrades.

Figure 18. Excerpt from typical specification. Another approach to the specifics of coordination, these pointed references leave little doubt as to what is required.

tion system looking over his shoulder almost constantly. Each of the functions described rests with the contractor, who is the end of the line. Simply put, it is the contractor who is responsible for the execution of the work. Therefore, the inspection system will be looking to him to execute the work properly.

The inspection system is not intended or designed to "come down on" the contractor. It imposes on him a system of checks and bal-

SGC-23   EXAMINATION:

a. Before starting the work, and from time to time as the work progresses, the Contractor and each subcontractor shall examine the work installed by others insofar as it applies to his work and contractor shall promptly notify the Architect if any condition exists that will prevent him or his subcontractor giving satisfactory results in his work. Should the work be started without such notifications, it shall place upon him the responsibility for replacing any of his work that it may be necessary to remove, in order to correct such faults.

Figure 19. Excerpt from typical specification. Called by another name, this is a requirement for inspection by all contractors involved.

PART 1 - STANDARD AIA GENERAL CONDITIONS: (CONT'D)

ARTICLE 4:

4.2.1: DELETE ENTIRE PARAGRAPH.

4.5.1: WHERE THE WORD "ARCHITECT" OCCURS CHANGE TO READ, "DEVELOPER" IN LINE 7. DELETE THE WORD "ARCHITECT" IN LINE 2.

4.7.1: DELETE PARAGRAPH AS WRITTEN AND ADD IN ITS PLACE THE FOLLOWING: "THE GENERAL CONTRACTOR SHALL SECURE AND PAY FOR ALL PERMITS, UNLESS MODIFIED BY THE DEVELOPER-GENERAL CONTRACTOR AGREEMENT, INCLUDING GENERAL WORK BUILDING PERMIT AND OCCUPANCY PERMIT. GOVERNMENTAL FEES AND LICENSES, INSPECTIONS, AND ALL OTHER LEGAL FEES PERTAINING TO HIS TRADE BOTH PERMANENT AND TEMPORARY WHICH ARE NECESSARY FOR THE PROPER EXECUTION AND COMPLETION OF THE WORK, WHICH ARE APPLICABLE TO THE CONSTRUCTION PROCESS."

4.7.2: DELETE PARAGRAPH AS WRITTEN AND ADD IN ITS PLACE THE FOLLOWING: "THE GENERAL CONTRACTOR SHALL GIVE ALL NOTICES AND COMPLY WITH ALL LAWS, ORDINANCES, RULES, REGULATIONS, AND ORDERS OF ANY PUBLIC AUTHORITY BEARING ON THE PERFORMANCE OF THE WORK. IF THE GENERAL CONTRACTOR OBSERVES THAT ANY OF THE CONTRACT DOCUMENTS ARE AT VARIANCE THEREWITH IN ANY RESPECT, HE SHALL PROMPTLY NOTIFY THE DEVELOPER, ARCHITECT, IN WRITING, AND ANY NECESSARY CHANGES SHALL BE ADJUSTED BY APPROPRIATE MODIFICATION. IF THE GENERAL CONTRACTOR PERFORMS ANY WORK KNOWING IT TO BE CONTRARY TO SUCH LAWS, ORDINANCES, RULES, AND REGULATIONS, AND WITHOUT SUCH NOTICE TO THE DEVELOPER AND ARCHITECT, HE SHALL ASSUME FULL RESPONSIBILITY THEREFOR AND SHALL BEAR ALL COSTS ATTRIBUTABLE THERETO."

Figure 20. Excerpt from typical specification. Supplements to the *General Conditions* modify the basic requirements for the contractor and others. These are usually imposed for special requirements of the individual project.

40 Systematic Construction Inspection

ances in that he has to meet the criteria of various regulations and agencies. A contractor can spend a great deal of his time involved with the inspection process, and much of his involvement can be traced directly to his philosophy. If he chooses to be a straightforward, cooperative contractor, demanding the best work of himself, his staff, and his subcontractors, he will usually have very few problems with the inspection system. If, however, he chooses to take shortcuts with an overaggressive approach that results in doing the wrong thing at the wrong time, if he is constantly in an adversary position, if he carries a chip on his shoulder from other jobs (with some of the same participants), he can very well be in trouble with the inspection system.

The system is not designed to be punitive; it is not "after" any particular person. It is there to ensure that the work is executed in a proper manner. The contractor, by bidding on the project and signing the contract, has voluntarily put himself in the position of responsibility. He has not been forced or cajoled into the position; he has accepted the contract openly and knowingly. He knows what procedures are necessary, and his philosophy will guide him in dealing with each element of the project.

It would be very helpful for the contractor, just as it is for the design professional, to seek out all the inspection agencies on the job and to list them, thus having available quick reference to the proper people, telephone numbers, addresses, and so on. It would be well to have not only this list, but some notes about the requirements as well, such as whether notice prior to inspection is necessary, when that notice must be given, what the particular inspector is looking for, or at what time inspection is required. It is far easier for the contractor if he can predict what the inspection system will require and when the requirements will engage the project. Planning for the inspections is very wise, in that the work can be finished and examined beforehand, job sequence need not be disrupted, and proper scheduling can be maintained. Also, any faulty work can be found early and promptly remedied.

RESOLVING INFORMATION FROM OTHERS

The contractor, while participating in the inspection process, is placed in the position of gathering a great deal of information from other sources. This discussion deals only with the other inspection sources that are imposed on the project. Various people will be inspecting

---

PERMITS AND FEES

Contractor shall secure and pay for, all permits and any fees required for carrying out the work, as may be required by governing authorities. Each contractor shall obtain and hold in force, any item required to complete that portion of the work. The owner shall obtain and pay for general building permit.

CODE COMPLIANCE

The Contractor shall give all requisite notices to the proper authorities, obtain all official inspections, permits and licenses made necessary by the work and shall comply with all laws, ordinances, rules and regulations pertaining thereto.

Figure 21. Excerpt from typical specification. The beginning of the inspection process. A requirement to meet the code with payment by owner for complying work.

the many aspects of the job at different times. It would be helpful if the inspection team and all its participants could move about the job periodically, side by side in a joint effort, but this is not always possible.

Although large groups may prove cumbersome and ineffective, several inspectors can review the job at the same time, allowing immediate decisions to be made and adjustments to be documented by all present. Preferably, such groups should include the design professional, the construction manager, and the contractor. Any such group should proceed in a controlled fashion and should not be allowed to become a "grand tour" of the project with little if any inspection other than some cursory viewing.

The contractor, then, with a myriad of information from various sources must begin to sort out all this information. It is possible that this information could be contradictory, out of sequence, possibly be out of date, or completely unfounded. A participant in the inspection system may very well try to force the contractor to execute some work not in keeping with the contract documents, but in the way that the participant wants it done, whether it meets the documents or the minimum standards or not.

Here the contractor must be not only astute, but also firm and flexible. He must be able to stand his ground and say "this is not in keeping with my contract; therefore, there must be some adjustment." If he is played the fool too often, the contractor can become very upset and require that certain conditions be met before he continues work. These conditions could be that all information must be in writing and verified by others or otherwise properly documented. It is far easier, however, if all participants recognize where their responsibilities stop and stay within them. This is particularly true with the requirements of the contract documents, which really are the parameters of the entire inspection system.

1.08 REQUIREMENTS OF REGULATOR AGENCIES:

    A. It shall be the responsibility of each contractor or subcontractor to apply for and obtain any and all permits and inspections which may be required for his work by local laws, ordinances, rules and regulations.

    B. Copies of all such permits and inspection certificates shall be filed with the Architect.

    C. The fees for such permits and inspections will be paid for by the contractor, securing same.

1.09 INSPECTIONS BY GOVERNING AGENCIES:

    A. Contractors for plumbing, heating, ventilation and air conditioning electric, and for any other work requiring special inspection, shall arrange for the inspection and test of the installation as required by the governing authority and/or by the specifications and shall provide all necessary tools, equipment, and personnel to conduct the required tests and shall notify the Architect at least three days in advance of such scheduled inspections and tests and shall submit approved certificate of inspection or copy thereof, from the governing agency prior to request for final payment.

Figure 22. Excerpt from typical specification. These provisions incorporate governmental inspection into the total project.

42 Systematic Construction Inspection

The contractor may have to do some fancy footwork from time to time in an effort to resolve or placate the various inspection sources. Again, he must be able to come forth, show the contradictory elements that are involved, and ask for guidance on how properly to proceed. In this event, the contractor should see this informational boondoggle as symptom of problems on the project. He may have to keep a constant watch for false, irrelevant and contradictory information, or else the problem may recur. If he has to deal with it on several occasions, he should take a very hard line with the inspection system and call it to task so that it performs properly. Proper communication should be reestablished, and the contractor and his project should not be placed in an adverse position.

ENSURING GOOD WORK

One of the major features of the contractor's inspection is to ensure that all work is properly executed; that is, he must ensure quality control. The specifications require a certain level of workmanship on almost every aspect of the work. The specifications should not be ambiguous; they should be precise and enumerate exactly what is expected: the quality of the work, the finished appearance of the work, and the workmanlike approach for installing the work.

The contractor should be aware of these job requirements. He must, therefore, be actively engaged in an inspection system that views all of them objectively. If something is not done in compliance with the contract documents, the contractor should correct it on his own initiative. Faulty work should not be ignored or allowed to remain uncorrected until another part of the inspection system catches it and takes action.

Many contractors, though, do not want to inspect the work before the more authoritative inspections. It may be a matter of time or priority, but more likely than not the contractors simply do not want to criticize the work of others. They feel the subcontractors know their work and should perform it properly; if they don't, then any remedial work or slowdown in job progress rests with the subcontractors. In the end, the project suffers from this lack of initiative, no matter who is responsible. All contractors should be made fully aware of their contractual requirements and should be made to perform accordingly.

A good contractor on a medium-to-large job can best serve himself with a good superintendent, preferably a nonworking one whose job is only administration and coordination. While this is another overhead cost that the contractor must absorb, it can enhance the profit of the contractor. A carpenter or carpenter-foreman who is required to produce a full share of the work himself cannot really be a proper overall superintendent of a construction job. Hence, he serves neither his employer nor the project well. A proper superintendent must be able to move around the job, checking on all phases of the work as it progresses. He should be intimately aware, of course, of all the contract document requirements, and he should be

---

1.02   WORK COORDINATION & JOB MEETINGS

    A.   Coordination: The Contractor(s) and subcontractor(s) shall, when so required to expedite the general progress of job, temporarily omit or leave unfinished any part of their work in order to facilitate work of others and Owner shall not be charged for such accommodations. Job Meetings every ten working days, or when otherwise scheduled by Architect.

Figure 23.   Excerpt from typical specification. This is a good requirement for all contractors to communicate and participate in discussions about the job.

an active inspector. He should have direct communication not only with the contractor's staff, but with all subcontractors.

This type of inspection should be ongoing from the first day of the project until the very last item on the punchlist is completed. The contractor can serve himself best by actively and aggressively participating in this quality control inspection. He can save himself a lot of headaches by looking at the work and having it corrected if improperly done. Of course, if there is any question in his mind as to the propriety of the work, he can call in any of the other participants to help him make a decision.

Too often with the tight scheduling of present-day construction projects, work is done and forgotten. It is never really inspected for quality control, and unfortunately too often the attitude of "Oh, it will be okay if nobody else catches it" is taken.

The inspection system is by no means perfect, and not all faulty work can be caught. It certainly is refreshing to have the contractor on the job constantly looking at the work to ensure that it is done properly. Such a procedure can ease the finalization of the project, and callbacks to the project will be minimized because the work was done properly in the first place.

CONTRACTOR'S SUPERINTENDENCE

The contractor shall keep the same superintendent on the job during its duration. Superintendent shall not be required to work with tools, or perform any other work not related to administering, expediting or coordinating the work under this contract. He shall have previous experience in this type of work and shall maintain progress schedule and be authorized to make field decisions in the absence of the contractor.

SGC-33   REQUIREMENTS FOR ALL CONTRACTORS:

   a. The Contractor shall keep on his work, during its progress, a competent Superintendent and any necessary assistants, all satisfactory to the Owner. The Superintendent shall not be changed except with the consent of the Owner, unless the Superintendent proves to be unsatisfactory to the Contractor and ceases to be in his employ. The Superintendent shall represent the Contractor in his absence, and all direction given to him shall be as binding as if given to the Contractor. Other directions shall be confirmed on written request in each case.

   b. The Contractor shall give efficient supervision to the work using his best skill and attention. He shall carefully study and compare all drawings, specifications and other instructions and shall at once report to the Architect any error, inconsistency or omission which may be discovered.

Figure 24. Excerpt from typical specification. This is an excellent statement about job superintendents and the work to be performed, requiring proper and constant inspection.

COORDINATION AND RESPONSIBILITY FOR SUBCONTRACTORS

On a traditionally organized project, the general contractor will have several subcontractors directly under contract to him. These subcontractors have no direct dealings with the owner, all of their business affairs being handled through the general contractor. Such subcontractors submit their bids to the general contractor, who reviews and analyzes them. Usually the lowest subcontractor's bid will be included as the price for that particular work in the overall bid submitted to the design professional and the owner. The contractual lines of the subcontractor system make it mandatory that the general contractor be responsible for and coordinate all his "subs," no one else has any direct leverage.

Although general contractors look for low bids, they also try to work with certain subcontractors. From experience, they know which subs are reliable, businesslike, and good mechanics. However, a formal contract should be drawn each time a subcontractor is given work. The benefits of using the better subcontractor should not be allowed to disappear because of informalities or uncertainties in contractual relations. Even good friends (contractors) should have businesslike contracts.

The general contractor is often confounded by the subcontractor, and will throw up his arms in frustration and lament that there is nothing he can do, that the subcontractor is a stubborn so-and-so and will not do his work as specified. The shrug and the lament do not, however, relieve the general contractor of his responsibility. Controlling the subcontractors is part of the general contractor's contractual obligations, and he cannot ignore it. He has contracted to provide that particular work, and he alone has chosen the subcontractor.

This responsibility is not simply to have the telephone number and a contract in the subcontractor's office; rather, a good working relationship is the intention of the contract. The two parties must work together and must schedule their work properly. The subcontractor should understand that his work, first of all, will be scrutinized by the contractor. The contractor should never feel that he's being intimidated by the subcontractors. He should retain control and ensure that all work is proper before the sub is in any way released from the project or to another work area.

The general contractor needs a good working knowledge of all phases of the construction process. He knows what is required from the drawings and specifications, and he has the first-line quality control over the subcontractor. This calls for inspection, plain and simple. The contractor should be willing to take on this inspection, and the subcontractor should understand that this is the first of several inspections to which he will be subjected.

Inspection should never be punitive. It should be a process of ensuring that the work is done correctly and completely and left in good order, meeting the requirements of the contract documents. The general contractor should be an active and willing participant, because if the work is not correct, it will be the general contractor who will be called to account. He, in turn, will have to try to deal with the subcontractor. A proper relationship and inspection at the moment the work is completed can lead to good feelings, respect, and an enjoyable project atmostphere. The alternative can drastically effect the project and the future relationship (and reputation) of both the general contractor and the subcontractor.

In a system where the general contractor no longer is the main functionary, the construction manager is responsible for the subcontractors or, more properly stated, for the various contractors. The manager must participate in the inspection sequence to ensure that the work is ready for inspection by others within the system. If this is carried out, it greatly expedites the work and can substantially reduce the amount of inspection time needed on the job. One of the greatest by-products of the system is, again, the elimination of hard feelings and of extended periods of time to have corrections made. It is far easier for the contractor or the subcontractor to replace or repair work while he is still on the job site than to have to come back to the job for a few hours' work to make an adjustment. Not only does this expedite the work, but it also reduces the cost to almost everyone involved.

TEMPO OF THE JOB

The general contractor, or the construction manager, has an unwritten responsibility that involves the general handling or the tempo of the project. Nothing is written into the contract, however, that says the job must be run in a certain way. The contractor, through his years of experience, will know that a well-run, well-scheduled, well-inspected project will progress very well. For the most part, it will seem to run itself, and, in the final analysis, it will be a well-executed project.

Many people can work very hard on the tempo of the job, and yet some can eminently succeed and others can fail miserably. There are no real guidelines on how to run a job successfully. Good management practices, of course, are necessary as well as a good knowledge of the construction industry, of current labor and economic conditions within the community, and of the system under which the project is being worked.

The inspection system plays a part in the running of the job. If the system is allowed to deteriorate and becomes a matter of "I'm going to do the work, catch me if you can," or if it becomes full of conflict, the tempo of the job is going to be slowed. It's going to be erratic, jerky, and generally unproductive. With this in mind, the contractor is best advised to take the inspection system into account and anticipate it at every turn. In this way, the inspection system becomes a known quantity.

The inspection system is not intended to place project members constantly on the defensive. It is a simple matter of maintaining the checks and balances. If the job becomes notorious because of its failures, the inspection system is going to have a more direct and frequent influence on the project. If the job is well executed, well run, and coordinated with the requirements of the contract, the inspection system will become a fringe system doing its job, but in no way impinging on the jobs progress.

The tempo of the job is set by the contractor or the construction manager in the initial stages. A cooperative spirit, an all-encompassing communication system, an openness, and a willingness to listen—all contribute to the tempo of the job and the successful incorporation of the inspection system. To be devious or to try to ignore the inspection system will lead only to the downfall of the general progress of the job. The system exists to satisfy certain requirements, but it should never be allowed to become so important that the job suffers. However, if the job is poorly executed from the beginning, the inspection system will have a very marked impact on the job. There is no implied threat in this. It is a matter of evaluating the work against job and code requirements.

The great majority of construction projects are well-run and well-executed, and all participants walk away friends, proud of their joint accomplishment. There is no reason this can't happen on every job. No inspector, be he a design professional, an owner, or from a governmental agency, is on site to harrass the contractor, impede the job, inflate his own ego, or seek his own self-interest. He is there for the good of the project. The inspection system itself can, for the most part, be self-enforcing. One element of that system will not allow another element to become too pushy or to impinge on progress.

Someone, though, must start the ball rolling, and this responsibility falls on the contractor or the manager. The job ideally begins with proper scheduling, proper communications, and an allowance for the inspection system. This allowance does not mean that on certain days the entire job will just shut down so everything can be inspected. This means not only that the inspection system is acknowledged, but also that it will be called in when required, that work will be ready for inspection when required, and that the work will have had some preliminary inspection prior to the major inspection.

With this sequence, the inspection system will add to the tempo and help the job. It will stay out of the way, allowing progress and a well-executed job, meeting the budget and schedule without any problems.

CHAPTER 6
THE SUBCONTRACTOR

SUBCONTRACTOR *n.* A secondary contractor who performs some part of the prime contractor's obligation under the contract.

CONTRACTUAL OBLIGATIONS

The subcontractor on a construction project is contractually the "end of the line." He is basically responsible for a specific, limited amount of the work to be done on the project. Although the scope of his work can vary greatly, his interest in many ways is extremely narrow.

In the current system, the subcontractor may indeed subdivide his work and have a series of sub-subcontractors below him; even further subdivision is possible. In addition, there are contractual agreements with material suppliers, manufacturers, distributors, and manufacturer's representatives. This entire group of participants in the construction process, however, can be categorized under the general heading of subcontractor.

All these participants have the same basic contractual obligations and responsibilities. Each has a limited, narrow responsibility to the overall project. His responsibility lies in having a smaller organization than that required to produce the entire project, and his work is highly specialized. He could be in charge of an aspect as large as the entire masonry or bricklaying contract, or he might provide only a portion of the carpeting for the project and nothing more.

Basically, the subcontractor provides the work force, the actual hands-on work force, for the project. The administration of the subcontractor is fairly small and is confined to the overall administration of the business and work under contract. The subcontractor has only a small portion of the administrative responsibilities on a large project. He must keep in touch with each project to know when his work is required and what time schedules he must meet. Although he may be only a small cog in the machinery of a large project, he is nonetheless very important to the overall quality, progress, and completion of the project.

All of the requirements he must meet should be enumerated in the contract the subcontractor has with the general contractor. Of course, the general contractor, having overall administration, inspection facilities, and functions on the project, will be looking very carefully and objectively at the subcontractor, wanting to ensure that the subcontractor performs well and that the work will not be a source of either irritation or callbacks on the part of the contractor. Also, he will want to see that the subcontractor performs in a timely manner—getting onto the job, doing the work, and then leaving in a expeditious manner when the work is finished.

RESPONSIBILITY TO EVERY MEMBER OF THE DESIGN/BUILD TEAM

It is important that each subcontractor feels that he is a part of the project. He should be included in the job meetings when he is on the

6. COMPLIANCE WITH LAWS:

A. EACH SUBCONTRACTOR SHALL ASSUME ALL RESPONSIBILITY AND HOLD THE ARCHITECT AND THE DEVELOPER HARMLESS FOR COMPLIANCE WITH ALL APPLICABLE LAWS, ORDINANCES, RULES, REGULATIONS, AND ORDERS OF ANY PUBLIC AUTHORITY BEARING ON THE PERFORMANCE OF THE WORK, INCLUDING FURNISHING AND INSTALLING OF ALL LABOR AND MATERIALS REQUIRED TO OBTAIN THE OCCUPANCY PERMIT.

B. EACH SUBCONTRACTOR, INCLUDING ALL EMPLOYEES, SHALL COMPLY WITH OSHA TO THE EXTENT REQUIRED BY THE DEPARTMENT OF LABOR, OCCUPATIONAL SAFETY AND HEALTH ADMINISTRATION. THE GENERAL CONTRACTOR SHALL NOTIFY THE DEVELOPER AND THE ARCHITECT OF ANY CHANGES REQUIRED BY OSHA.

C. EACH SUBCONTRACTOR WARRANTS THAT ALL CONSTRUCTION AND INSTALLATION WORK SHALL BE PERFORMED IN ACCORDANCE WITH THE REQUIREMENTS, ORDERS, AND LIMITATIONS OF ALL LOCAL, STATE, OR FEDERAL DEPARTMENTS OR BUREAUS HAVING JURISDICTION. UPON COMPLETION, THE PREMISES SHALL BE IN COMPLIANCE WITH ALL GOVERNMENTAL REQUIREMENTS FOR THE USE WHICH THE DEVELOPER MAY MAKE OF THEM, AND THAT ALL NECESSARY CERTIFICATES OF INSPECTION SHALL BE OBTAINED BY THE GENERAL CONTRACTOR AND SUBMITTED TO THE DEVELOPER.

D. EACH SUBCONTRACTOR SHALL CAREFULLY INSPECT ALL DIVISIONS OF THESE SPECIFICATIONS, AND THE CONSTRUCTION DRAWINGS, PRIOR TO THE SUBMISSION OF HIS BID, AND NOTIFY THE GENERAL CONTRACTOR OF ANY NON-COMPLYING MATERIALS OR ERECTION METHODS. IF THE SUBCONTRACTOR FURNISHES ANY WORK WHICH IS NOT IN CONFORMANCE WITH SUCH LAW, ORDINANCES, RULES, AND REGULATIONS, AND WITHOUT NOTICE TO THE GENERAL CONTRACTOR, HE SHALL BEAR ALL COSTS ARISING FROM THE CORRECTION THEREOF.

E. ALL PERMITS, LICENSES, CERTIFICATES AND NECESSARY INSURANCE REQUIRED FOR THE WORK SHALL BE OBTAINED AND PAID FOR BY THE SUBCONTRACTOR, AS HEREINAFTER SPECIFIED AND IN OTHER CONTRACT DOCUMENTS.

F. ALL ELECTRICAL EQUIPMENT AND INSTALLATIONS SHALL BE APPROVED BY THE UNDERWRITER'S BOARD AND LOCAL ELECTRIC COMPANY; ALL GAS EQUIPMENT AND INSTALLATIONS SHALL BE APPROVED BY THE LOCAL GAS COMPANY.

G. COMPLIANCE WITH LAWS, ETC., SHALL NOT BE USED AS A MEANS OF JUSTIFYING THE INSTALLATION OR APPLICATION OF PARTS, ASSEMBLIES OR METHODS INFERIOR TO THOSE SPECIFIED.

Figure 25. Excerpt from typical specification. A listing of all requirements imposed on the subcontractor implies inspection in several instances.

job and performing his work. If he is ignored, he may feel that he is merely putting in his time and that he owes no responsibility to the overall project. In fact, he should never feel that he is being ignored, bypassed, or taken for granted; rather he should be respected for the knowledge and ability that he brings to the project. His responsibility to each member of the design/build team, of course, is to produce the best work he possibly can, utilizing the best mechanics and workmen and always using the material and methods listed in the specifications for the project.

The subcontractor is also a full participant in the inspection system. He should determine that the standards applicable to his work are fully met and that all work is installed in keeping with the manufacturer's printed instructions.

1.06 ADHERENCE TO CODES AND REGULATIONS:

A. Before proceeding with the work, each contractor shall thoroughly review the drawings and the specifications to assure the design is in accordance with all laws, ordinances, regulations, and building code regulations and building code regulations applicable to work hereunder.

B. Should a discrepancy to such applicable laws, ordinances, codes or regulations be determined, the discrepancy shall be brough to the attention of the Architect who will provide corrective instructions.

C. If the contractor performs any work knowing it to be contrary to applicable laws, ordinances, rules and regulations, and without notice to the Architect he shall assume full responsibility therefore and shall bear all costs attributable thereto.

D. The "Ohio Building Code" as administered and modified by the Division of Public Works, City of Cincinnati, Ohio shall govern all the work in addition to any other code authority indicated in specifications.

1.07 APPLICABLE STANDARDS:

A. Where reference is made in the specifications to published standards, the issue of such standard current on date of invitation to bid shall govern the applicable work.

B. In case of conflict between the published standard and project specifications, the latter shall govern.

C. References to known stadnard specifications shall mean the latest edition of such specifications adopted and published at date of invitation bids. Reference to technical society, organization or body is made in specifications in accordance with the following abbreviations:

Figure 26. Excerpt from typical specification. Contractor requirements insure that the project meets all applicable laws and codes, ensures understanding of the project, and allows input from the contractor, who may have information different from that of the design professional.

GENERAL GYPSUM BOARD INSTALLATION REQUIREMENTS:

A. Pre-Installation Conference: Meet at the project site with the installers of related work and review the coordination and sequencing of work to ensure that everything to be concealed by gypsum drywall has been accomplished, and that chases, access panels, openings, supplementary framing and blocking and similar provisions have been completed.

Figure 27. Excerpt from typical specification. This excellent requirement brings the subcontractor into the project. It coordinates his work and helps him understand the project as a whole.

His work force should be trained in the minute detail of all the products they provide and install, but inspection of the work is still essential.

By performing in this manner, the subcontractor will have a long business life, in that he will be treated with respect and will gain in stature as he takes on more responsibility. He will distinguish himself as he performs in a good, and workmanlike manner, extending himself in the best way possible and keeping in mind that the project, of course, should be placed above all else.

However, the smart subcontractor performs only the work for which he is engaged, and he should not allow any outside influence to sway him from his work or to convince him to do work for which he is not trained or qualified.

This is not to say that the subcontractor should be constantly looking for loopholes or shortcuts. If he is a responsible businessman, he has some basic business sense. He will know that his contract price is a valid price and that his materials are as specified. He should then in a businesslike way be ready to perform

---

1.03 QUALITY ASSURANCE:

A. Job Mock-Up:

1. Prior to installation of masonry work, erect sample wall panel mock-up using materials, bond and joint tooling required for final work. Build mock-up at the site, where directed, of full thickness and approximately 4'x4', indicating the proposed range of color, texture and workmanship to be expected in the completed work. Obtain architect's acceptance of visual qualities of the mock-up before start of masonry work. Retain mock-up during construction as a standard for judging completed masonry work. Do not alter, move or destroy mock-up until work is completed. Use sample panels to test proposed cleaning procedures. Provide separate mock-up for the following:

   a. Typical face brick wall.

Figure 28. Excerpt from typical specification. This requirement establishes a model against which other work can be evaluated. The model should be executed in keeping with all standards, for it is easier to change a faulty model than a large portion of the work in place.

PROTECTION OF HIS RIGHTS, INTERESTS, AND BUSINESS

The inspection aspect of the subcontractor's work involves more self-interest than that of any other inspection participant. His inspection should be aimed directly at his own interests to ensure that the quality of work he is producing meets the specifications and the requirements of the project. At the same time, it should not become burdensome to his own work.

Being the "bottom-line, do-it" contractor, the burden to perform weighs heavily on the sub.

to the best of his ability. No one is asking any more, and no less should be expected of him. His inspection should involve his self-interest in that he should look at the work of others in relationship to his own. For example, a contractor under contract to build the structural frame of a building should investigate and inspect the foundation work that has already been done by another contractor. It would be foolhardy for him to begin work without doing so and find that problems with his work are created by faulty foundations.

Similarly, a flooring contractor should very carefully inspect the concrete floor slabs before he begins installing his material. Rough surfacing, cracks in the slab, or other problems with the concrete could cause problems for the subcontractor, costing him additional money to remedy and to ensure that his work is finished in a complete and proper manner. In most specifications the writer will specifically call for these inspections and place the responsibility for them on the subcontractor for the new work. In other words, the contractor for the structural

---

PART 3.00 - EXECUTION

3.01 INSPECTION:

A. Applicator must examine the areas and conditions under which painting work is to be applied and notify the Contractor in writing of conditions detrimental to the proper and timely completion of the work. Do not proceed with the work until unsatisfactory conditions have been corrected in a manner acceptable to the Applicator.

B. Starting of painting work will be construed as the Applicator's acceptance of the surfaces and conditions within any particular area.

---

3 01 <u>INSPECTION AND PREPARATION:</u>
Inspect all surfaces to be waterproofed and remove all material that may be detrimental to bonding of waterproofing   Report to Architects any conditions which are not satisfactory for waterproofing application Application of material constitutes acceptance of surface conditions by waterproofing contractor.

<u>PART 3 - EXECUTION</u>

3.01 INSPECTION:

A. Installer must examine the areas and conditions under which resilient flooring accessories are to be installed and notify the Contractor in writing of conditions detrimental to the proper and timely completion of the work. Do not proceed with the work until unsatisfactory conditions have been corrected in a manner acceptable to the Installer.

Figure 29. Excerpt from typical specification. Three examples of inspection by the subcontractor for his own interest properly applies where some work is contingent on other work.

frame is given the responsibility to find any faults that may exist in the foundation system. So, too, the flooring contractor is responsible for any faults in the concrete floor slabs.

If a subcontractor finds anything wrong with the work that preceded him, he should report it in writing, through the proper channels so that it can be remedied prior to the installation of any new work. This, of course, is the reasonable approach and this procedure should be documented in the subcontractor's contract, as well as in the construction documents.

Although this may seem like self-preservation on the part of the subcontractor, it is only right and proper that the responsibility for any faulty work should not be shifted from one contractor to another. Future litigation or problems that develop later with the user of the project can be extremely hard to resolve if it is difficult to trace the lines of responsibility. Hence a proper contract, proper documentation, and proper inspections are essential.

The duration of legal liability of subcontractors, and indeed of all contractors in this country, varies from state to state. In about a dozen states, the courts have struck down the limit of liability for defects in a project, so that in some states the limit that had been set at ten years now lasts as good as forever. About two-thirds of the states still retain laws that set definite limits, but a trend may be developing that would demand even closer inspection by every contractor, because of extended liability.

The subcontractor, having a smaller organization, is not in a position to financially assume extra work on the job. He is not in a position to have his own work force cover a large amount of faulty work. Therefore, he must be very aware of all aspects of the work he is concerned with.

The subcontractor must be sensitive to the job requirements and to his own position. He should resist pressures that try to involve him in work unfamiliar to him or that is beyond his expertise and thus extends his liability. He must be aggressive but reasonable in his approach, doing his work properly and watching every condition around him.

His inspection carries over into his own work. He should be concerned that his forces are, first of all, meeting the requirements of the specifications. Of course, he will be inspecting his work to see that his workmen are performing according to his time schedule and are not extending his contract exposure time on the project to the detriment of both his finances and his reputation.

Ideally, every subcontractor would like to get the job, perform his specialty without interruption, leave the site, and get paid. Unfortunately, not all jobs proceed in this smooth sequence. Unforeseen time delays, return trips, long waiting periods, and the like, can lead the subcontractor to ruin.

A good contractor can very well get in, do the work, and get out without being detrimental to the project. If he has good workmen, has

---

C. Fire-Rated Assemblies:

1. Wherever a fire-resistant classification is shown or scheduled for hollow metal work, provide fire-rated hollow metal doors and frames investigated and tested as a fire door assembly, complete with type of fire door hardware to be used. Identify each fire door and frame with recognized testing laboratory labels, indicating applicable fire rating of both door and frame.

2. Construct assemblies to comply with NFPA Standard No. 80, and as herein specified.

Figure 30. Excerpt from typical specification. Detailed information, required to ease the burden of the inspector, requires code-abiding data to be displayed on the work.

his material on hand, and has a well-coordinated and well-administered organization, he should be able to perform in this manner and provide the very best work possible.

His inspection, of course, is specialized to whatever materials or systems he is working on, and he should be well trained in the inspection of those materials and systems. He should know his work literally inside out and should prepare it to meet the requirements of the job. It may well be that a subcontractor working on the job in a relatively late stage can be asked to do additional work to cover the mistakes of others. This, of course, should be added to his contract. He should not be forced to do this work without extra remuneration.

The subcontractor's inspection of work done previously should always be done in a spirit of cooperation and in the best interest of the project. The subcontractor should never be made to feel that he is betraying his fellow contractors by reporting work that was not performed properly. Unless the subcontractor's work is extremely limited in scope, it is awfully hard for every minute detail of every piece of work to be properly examined and inspected. He should inspect his work as others inspect theirs. Overall, though, he will allot a rather limited time for inspection. His people working in a limited area should be intimately aware of the details of this work. Some follow-up inspection is essential to see that the work fits properly into the project as a whole. Many times things will slip through and affect other contractors.

Figure 31. A poorly executed concrete job; note honeycombing, roughly framed opening, and structural fracture. This does not make a satisfactory base for any other work.

The spirit of cooperation is necessary among all participants, from the prime contractor to the material man who supplies the smallest item on the project. If this is not inherent in the project, there could be serious problems, and some people will try to avoid direct responsibility. This is another of the basic reasons for coordinated inspection system.

1.02 REQUIREMENTS OF REGULATORY AGENCIES:

a. All work included under this Project Manual shall conform to the National Electric Code as approved by the Inspection Bureau, Inc. and regulations of the Cincinnati Bell Company and the Cincinnati Gas and Electric Company. Include requirements of all codes whether shown on Drawings or not. Differences between the Bid Documents and local codes and regulations shall be corrected, as required, without any cost to Owner.

b. The Contractor shall obtain all required permits, temporary releases, inspections and include the cost in his Bid. Certificates of approval shall be submitted prior to request for final payment.

Figure 32. Excerpt from typical specification. Detailed requirements for the mechanical subcontractors state that all work on the project must comply with all applicable regulations.

CHAPTER 7
THE BUILDING INSPECTOR

BUILDING INSPECTOR *n.* A representative of a governmental authority employed to inspect construction for compliance with applicable codes, regulations, and ordinances.

BACKGROUND

Building codes were born of disaster, human tragedies, and violations of public health, safety, and welfare. For over 5,000 years, governmental officials have found it necessary to control the construction and placement of buildings within civilized communities. As soon as the human race found it necessary to construct shelter for protection from the elements, hazardous situations were encountered. The lack of knowledge and information about natural laws and the capabilities of various materials that could be used for building were initially a danger. Earthen walls and roofs, as well as light wood structures, could collapse because the most basic engineering requirements and principles were not known. The only "method" was that of trial and error, which could easily lead to disaster. Even when fire was controlled, providing warmth and comfort, it could also be a tremendous hazard to the user's well-being. This is the basic root of the building code and inspection system that exists today.

The system of construction now enforced is a tug-of-war between the best interests of the owner and developer and the best interests of the general public. The best interests of the owner is not always, and is not necessarily, in the best interest of the public, and it is this that is responsible for the imposition of governmental inspection on construction projects. Governmental inspection includes building inspection, inspections by fire-prevention officers, health and sanitation officals, plumbing inspectors, electrical inspectors, and inspectors of the various other engineering diciplines. Here, the building inspector is discussed primarily, although all governmental inspectors act in basically the same way.

Figure 33. The building inspector and the fire service inspector coordinating a joint inspection. All requirements can then be met with one inspection.

Simply put, construction drawings and specifications are a commitment to comply with the standing law of the jurisdiction, the building code. The building inspector and his inspection system is a check mechanism to verify that such compliance is achieved. It could very well be that the owner or the use of the building will, at some time, imperil the building and its occupants. It is imperative, then, that the building inspector, speaking for the governmental jurisdiction, see that the finished building turned over to the using public is constructed to abide by the code as closely as possible.

The use of the building must be begun with the code in mind in order to prevent future misuse, the downgrading of protection, and the addition of fire loading to the structure. But here the interests of the owner and developer can directly conflict with those of the public. The owner, by virtue of his financial commitment, feels very deeply that he has a right to develop and use his building in his best interests and in the way that he wants to use it. But just as any right has a definite limit, so too this right of use has a limit, that limit being the imposition of the best interests of public health, safety, and welfare.

Well-intentioned, expensive construction assemblies frequently are rendered ineffective by lack of attention to significant detail in the design process, poor execution of the work, lack of proper in-progress inspection, or improper renovation after final inspection. These include the following.

Within buildings—

Improper or lack of fire separations (vertical and/or horizontal).

Lack of proper protection for such floor and wall openings as stairs, doors, windows, pipe penetrations, shafts, ducts, chutes, conveyors, elevators, dumbwaiters, and escalators.

Inadequate fire-stopping, or divisioning in concealed spaces above ceilings or in walls.

Application, or installation of combustible interior finishes, protective coatings, and insulations.

Fire walls supporting combustible framing members, such as joists, beams, and girders.

Poor anchorage of structural members into masonry walls.

Improper explosion venting.

Improper venting facilities for fire gases.

Violation, or improper construction of floors, and walls installed to act as fire barriers.

Between buildings—

Lack of adequate fire separation distance.

Inadequate fire resistance of exterior walls.

Improper opening protection.

Lack of adequate fire walls between buildings.

Combustible construction and materials; roofs, roof coverings, roof structures, overhanging eaves, trim, and the like.

Inadequate protection at connections between buildings; utility tunnels, passageways, ducts, conveying systems.

Collapse of protective exterior walls due to explosion or fire progress.

In very few cases is there a direct attempt to circumvent the provisions of the building code, although some, of course, have planned their buildings to achieve something outside the code. But on the whole, the design professionals and the owners want their buildings to abide by the code and the law.

The problem today is that neither the owner nor the design professional wants to take the time to explore and understand the details of the code or the rationale behind its development. The time essential for a proper design is key to both the design professional and to the owner. They are simply more interested in developing the structure and imposing the owner's program than they are in seeing that every detail of the building code is met. Neither the owner nor the design professional is trying to circumvent minimum standards; rather, there is such a massive array of information touching on the construction of a building that it is very hard to be familiar with all the details.

The building code and inspection system set up a program of review to see that the documents show compliance with the code, and the inspections verify that the construction itself abides by the code. This is the basic concern of the existing inspection system.

Although the design professional, the owner, the construction manager, the contractor, and the subcontractor all have an inspection function, they also all have a varying degree of self-interest in the project. Each has an axe to grind, which has a tendency to color his inspection capabilities and intentions. By contrast, the governmental inspector comes to the job with no self-interest. His basic charge is to check for compliance with the applicable codes. He is acting for the general public, and its interest is his prime mandate.

In years past the governmental inspector was suspect, mainly because of his credentials (or lack thereof), his tremendous power, his ability to influence a project by approval or nonapproval, and the manner by which he got his job. With more complicated projects and an enlightened electorate, the basic protection of the code officials system has caused the system to move toward a much more professional level. Still, the governmental inspector is the only truly impartial party on the project, his basic interest being the protection of the using public and not of any individual participant on the job. It is amazing that, with all the research done before a project is started, the code compliance inspection is needed more today than ever before. Designs are being produced in a very short time, and most shortcomings can be traced to hasty and inadequate document preparation. They are not being produced to purposely circumvent the code.

The design professional is not taught how to deal with codes, learning of such regulations only through his experience with them on the jobs, and has a tendency to ignore the more refined and involved details. He may have an intuitive feeling that he will be caught and that at that time he can remedy any violations of the building code. Moreover, he feels that his time is better spent attending to other parts of the project. He knows that his drawings and specifications will be reviewed by the building department prior to issuance of the building permit, and what is not caught in the way of code compliance at this juncture surely, he feels, will be caught during the inspection sequence.

Similarly, the contractor is not aware of all the details of all the basic reference standards contained within any code. The contractor has a tendency to work on a project in his own way and in the manner in which he was trained. His training may or may not have been in coordination with, or with the knowledge of, the minimum standards contained in the building code. How, then, is code compliance achieved?

Basically, information is gathered about various aspects of the project. Some are compared with code requirements. The plan review, prior to the issuance of a permit, usually reveals discrepancies with the code. The inspection system is in place to review work in progress to ensure compliance with the final approved contract documents.

There will always be a need for an inspection system, but it is becoming extremely important that code compliance become part of the basic design sequence. More and more design professionals are seeing that it is virtually impossible to separate the owner's design program criteria from the criteria of the building code. The building code contains massive, wide-ranging, and very complex regulations of a broad scope. Many consider these to be restraints, restrictions, or impositions that destroy the basic design concepts. This is not true. The design professional who is initially aware of the code provisions can, with the help of the building official, come to an understanding that will allow for accommodation of the design and still provide for basic compliance with the code. An increasing number of design professional offices are developing code search documents that are filled in during the initial design stages of each project. This code search provides the "public" requirements of the job. When coupled with the private owner's requirements for the job, the full parameters of the project are established.

It is a simple fact that the design professional is putting himself in the position of a baseball manager. No matter how many times the manager has had his team play in a certain baseball park, at home plate before every game there is a review of the ground rules for that park. Design professionals are beginning to realize that for

SECTION 00855 - STANDARDS AND ABBREVIATIONS

PART 1 - GENERAL

1.01 DESCRIPTION:

The following is a numeration of Standards and Abbreviations used in the Documents. Unless indicated, the latest published edition in effect during the bidding period shall govern the construction of this Project.

PART 2 - PRODUCTS/STANDARDS - ABBREVIATIONS

AA	Aluminum Association, Inc.
AAA	American Arbitration Association.
AABC	Associated Air Balance Council
AAMA	Architectural Aluminum Manufacturers' Association
AASHTO	American Association of State Highway and Transportation Officials
ABPA	Acoustical and Board Products Association
ACI	American Concrete Institute
ADC	Air Distribution Council
AFI	Air Filter Institute
AGA	American Gas Association
AGC	Associated General Contractors of America, Inc.
AHDGA	American Hot Dip Galvanizers Association, Inc.
AI	The Asphalt Institute
AIA	American Institute of Architects
A Insur. A	American Insurance Association
AISC	American Institute of Steel Construction, Inc.
AISI	American Iron and Steel Institute
AITC	American Institute of Timber Construction
AMCA	Air Moving and Conditioning Association
ANSI	American National Standards Institute
APA	American Plywood Association
AREA	American Railway Engineering Association
ARI	Air Conditioning and Refrigeration Institute
ASHRAE	American Society of Heating, Refrigeration and Air Conditioning Engineers
ASME	American Society of Mechanical Engineers
ASTM	American Society for Testing and Materials
AWI	Architectural Wood Work Institute
AWWA	American Water Works Association
AWS	American Welding Society
AWPA	American Wood Preservers Association
AWPB	American Wood Preservers Bureau
AWPI	American Wood Preservers Institute
BHMA	Builders Hardware Manufacturers Association
BIA	Brick Institute of America

Figure 34. Excerpt from typical specification. An elaborate listing of the many standards that apply to construction work provides basic, standardized, well-researched material. Most of these will appear in the building and other codes; others are imposed by the design professional.

BRI	Building Research Institute
BSI	Building Stone Institute
CISPI	Cast Iron Soil Pipe Institute
CLFMI	Chain Link Fence Manufacturers Institute
CPI	Clay Pipe Institute
CRA	California Redwood Association
CRI	Carpet and Rug Institute
CRSI	Concrete Reinforcing Steel Institute
CS	Commercial Standards, U.S. Department of Commerce
CSI	Construction Specifications Institute, Inc.
DHI	Door and Hardware Institute
EJMA	Expansion Joint Manufacturers Association
FM	Factory Mutual Engineering Corporation
FMA	Flexicore Manufacturers Association
FS	Federal Specifications
GA	Gypsum Association
GSA	General Service Administration
HPMA	Hardwood Plywood Manufacturers Association
IBRM	Institute of Boiler and Radiator Manufacturers
IEEE	Institute of Electrical and Electronic Engineers
IES	Illuminating Engineering Society
IPCEA	Insulated Power Cable Engineers Association
IRI	Industrial Risk Insurers (Formerly FIA)
LIA	Lead Industries Association, Inc.
MBMA	Metal Building Manufacturers Association
MIA	Marble Institute of America
MLSFA	Metal Lath/Steel Framing Association
NAAMM	National Association of Architectural Metal Manufacturers
NBHA	National Builders' Hardware Association
NBS	National Bureau of Standards (U.S. Department of Commerce)
NCMA	National Concrete Masonry Association
NCPI	National Clay Pipe Institute
NEBB	National Environmental Balancing Bureau
NEC	National Electrical Code
NEI	National Elevator Industry, Inc.
NEMA	National Electrical Manufacturers Association
NFPA	National Fire Protection Association
NMWIA	National Mineral Wood Insulation Association
NPA	National Particleboard Association
NTMA	National Terrazzo and Mosaic Association, Inc.
NSC	National Safety Council
NSF	National Sanitation Foundation
OSHA	Occupational Safety and Health Administration
PCA	Portland Cement Association
PCI	Prestressed Concrete Institute
PDI	Plumbing and Drainage Institute
PE	Perlite Institute
PEI	Porcelain Enamel Institute
PS	Product Standard of NBS (U.S. Department of Commerce)

Figure 34. (*Continued*)

RIS	Redwood Inspection Service (Grading Rules)
RTI	Resilient Tile Institute
SDI	Steel Deck Institute
S.D.I.	Steel Door Institute
SJI	Steel Joist Institute
SMACNA	Sheet Metal and Air Conditioning Contractors National Association, Inc.
SPI	Society of the Plastics Industry
SSPC	Steel Structures Painting Council
SWI	Steel Window Institute
TCA	Tile Council of America, Inc.
UL	Underwriters' Laboratories, Inc.
WMA	Wallcovering Manufacturers Association
WRI	Wire Reinforcement Institute

Figure 34. *(Continued)*

each project, no matter how similar it may be to previous projects, requires a reveiw of the "ground rules," that is, the basic code provisions.

Unfortunately, building codes currently being enforced contain long lists of reference standards. The codes would be impossibly large documents if all of the material contained in the reference standards were enumerated directly in the code. The standards are incorporated by reference only. Very few design professional firms, or even building departments, have all the standards immediately available to them in their offices. In fact, it is not very hard to find some standards contained in the codes are now out of print. These requirements must still be met, even though they are no longer readily available.

The design professional and the owner must come to the realization that once there is an invitation to the public to enter a structure, whether it is for a social visit or to do business, they both have a responsibility for the protection of the public. That protection begins when entrance to the structure is achieved, continues during the occupancy of the structure, and requires proper exiting. The responsibility is compounded when there is some sort of an emergency inside the building, for the public must be protected until everyone can safely exit the building.

An owner should realize, or be made to realize, that he may very well have to spend money that will not directly benefit his interests. If he intends to do business, he must follow a prudent, thorough, code-abiding sequence for the protection of his visitors and customers. Every year there is an alarming loss of life and injury due to fire, building collapse, people falling, and the like. Most cases are litigated, and in many instances massive payments are made to the survivors or the maimed.

An owner can be a very prudent and sharp businessman, but he becomes very suspect when he begins to misuse his building. If he has not permitted his structure to be designed and built properly in the first place, he becomes even more suspect when somebody begins to look at how prudently his entire building development was achieved.

That owners, developers, builders and design professionals do not always act for the good of the public necessitates governmental jurisdiction and is the basis of the building inspection department. So many factors are at play that some must act in the public interest—not to intrude or prohibit and not to restrict or impose, but to set down a prudent and cautious course to be followed so that the business will have a long life and anyone touched by that business will have a chance for a long life.

PARAMETERS OF HIS JOB

At the outset, it should be understood that the function of the building inspection department, the code administrator, and any building official or inspector is not one of design or construction. No properly concerned building official will delve into the basic design premise of the structure in any way, shape, or form. He must have at his command the knowledge of construction and the knowledge and the rationale of the code requirements. The best statements that he can make to a design professional with a problem are perhaps the suggestions of two or three solutions to the problem. But the code official should not dictate the solution. He should point out the area of distress, indicate which provisions apply, and what latitude the code allows in this situation. He may then go on to suggest various general solutions that he has seen elsewhere, or which come to mind.

Private citizens building their own homes will often try to extract information from the building official regarding the design and construction. Any official trying to act in a friendly, cooperative manner can easily fall victim to such a request. His best answer, given in a friendly, cheerful, suggestive manner is to have the owner seek the advice of a design professional or some tradesman who is familiar with the solution to the problem. Too often a solution may be offered, only to have the citizen become very dissatisfied later on and distort the facts when telling others, including the news media. The owner may say that he was "made to do it this way" by the building official, or that the official threatened that approval would be withheld if it wasn't done this way and with this material.

Too often in the past, building departments have been graftridden for one reason or another because people wanted to circumvent the code. Sometimes code officials were not professional and chose to seek remuneration in the wrong way by suggesting certain materials, certain salesmen, and so forth. These practices are rapidly disappearing as professionalism becomes more the key word in building inspection.

It is no longer possible for a building department to be founded or allowed to continue in an uninformed, "hip-pocket" context. Construction is too complex and hazards too great to permit merely a passing stab at code enforcement. Every jurisdiction, no matter how large or small, should demand the best from its building department. This is not popular work, but history shows time and again the tragic results of inconsistent, incomplete, or lackadaisical enforcement. Good programs must be developed, proper codes adopted, well-trained and well-informed staffs hired, and proper financial and moral support given by the citizens and the elected officials.

The seat of knowledge in any building department lies within the office organization. It is here that, in many jurisdictions, the plan examiners are required to be licensed professionals, and it is here that the library of standards and information is kept and updated. There is no way that any jurisdiction can possibly require its field inspectors to carry all of the standards information in their automobiles. The basic concept of construction inspection lies in the fact that plan review done in a methodical, systematic manner in an office surrounded with all available information will lead to the discovery of the major problems with code compliance. Plan reviewers can then, in conjunction with the design professional, remedy the situation and have the documents revised and redrawn as required so that the approved drawings will properly reflect code compliance for the project involved.

The field inspector then utilizes these approved documents during his numerous inspections. The documents should be either with the inspector or on the project every time the inspector is on the project, and his inspection of the job should check the actual, in-place construction against the approved drawings. Any deviations he finds should be resolved among the contractor, the design professional, and the plan examiner.

Although the field inspector is given latitude and discretion, it is done in a limited manner. The basic approval is given by the plan examiner. Any deviations or changes should be coordinated with the examiner and should be fully understood by everyone involved, and properly documented.

It is absolutely essential that a complete and accurate job history be kept and placed in the department's files when the project is finished. This requirement does not vary one iota from project to project. Just because one project is smaller than another does not mean that fewer problems can develop and there is less need for a proper history. These files are invaluable to the department and all parties to the project in the event of problems or litigation. Laws vary, but many departments must retain such files for ten years or longer. Many now use microfilm to reduce bulk.

60 Systematic Construction Inspection

THE NEED TO UNDERSTAND VARIOUS CONSTRUCTION SYSTEMS

Each year many thousands of new materials are introduced into the construction industry. The vast majority of these are well-researched, well-founded, well-engineered products, which have a valid place in the modern construction system. Some, however, are produced with little research; some may even be plagiarized copies of other material systems and be poorly manufactured. As long as there is a variance between the basic concept and the production of new materials, there will be a need for an inspection system. The prudent materials manufacturer, in an effort to increase the marketability of his product, will find the necessary funding to research the basic design of his product, to properly develop the product, and to have the product tested to prove its reliability. The testing of the basic virtues of the material is vitally important to the building official and should be just as important to each of the participants in the construction system.

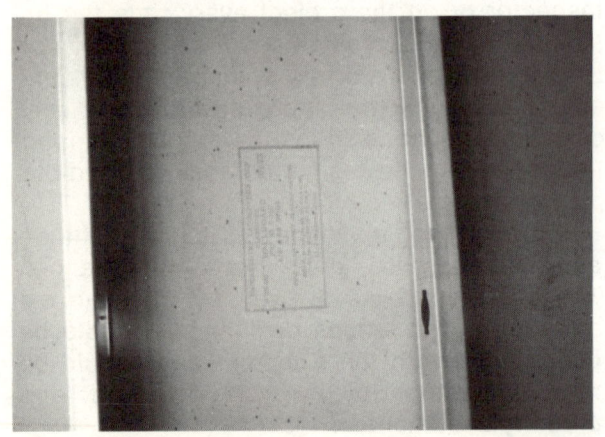

Figure 35. Fire-rating stamp displayed on the drywall board. This installation does not meet the fire resistance rating for the wall, just because the board is rated. The entire wall assembly must be in place for proper protection.

New material may appear on a project at any time. It is not the function of the building inspector or the building code to restrict and prohibit the introduction of new materials. It is an acknowledged right of the owner, contractor, or design professional to introduce what he considers to be a new alternative material. The problem becomes one of equivalency. To introduce a new material or a new system of construction into the project while it is in progress presents a problem to the building inspector. Does this material qualify as a direct replacement for the other material? Does it have the same attributes? Are the same virtues present whether they be structural, fire resistant, or decorative? The introduction of this material without data that substantiates equivalency becomes a problem. Careful, meticulous material selection takes place early in the design process, and the introduction of an alternative material that does not have the same attributes as the original begins to cause the deterioration of the project. If the particular product is to be fire-rated to a certain level and an alternative is introduced that does not carry the same rating, the project must be downgraded. Either the alternative must be prohibited or other adjustments must be made within the fire protection system to maintain the level required.

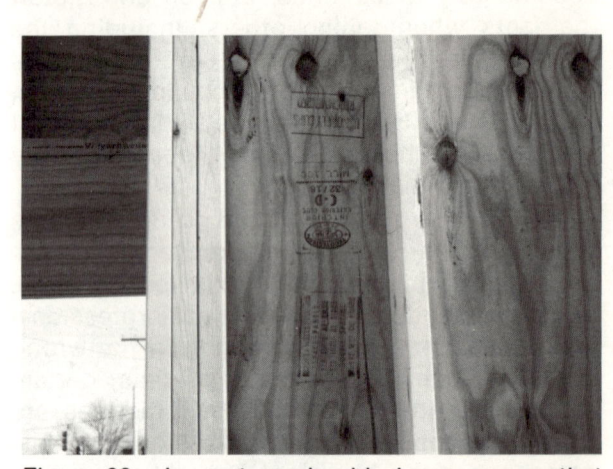

Figure 36. Inspectors should always ensure that all lumber displays the proper grade mark for the application involved. This code requirement is often overlooked.

How can all of this take place in the middle of a muddy construction site? The building inspector must know construction materials and systems at their basic, fundamental level. He cannot possibly carry with him the entire array of test results and the attributes of all materials and their alternatives.

The building inspector must be able to look at a particular system, and evaluate whether it meets the minimum standards of the building

Figure 37. Inadequate anchorage for the wall is unacceptable work.

Figure 38. Why continue to build the structure when such an obvious problem has occurred? The bearing is inadequate.

Figure 39. The bearing header is framed to double stud on one side, but why not on the other?

Figure 40. No toe-nailing shows the need for a flexible inspector. Codes usually don't recognize fastening other than that denoted in nailing schedules, but prefabrication of structures has imposed face nailing (which is acceptable).

code. Many times a building inspector, trained and active in a construction trade for many years, will have to swallow very hard to accept a piece of work. If he were to do the work, he would do it in an entirely different manner. But the building inspector must use the yardstick of the minimum standard in accepting or rejecting the work he sees. The inspector simply cannot impose his will or his method on the project, since neither of these may be in keeping with the minimum standard. He cannot ask for more than what the standard requires, but he certainly cannot tolerate less.

Prudence and flexibility are two tremendous attributes that the inspector must have. He must be strong enough to say, "Stop, we must investigate this situation." Often it's very hard to get a contractor to cooperate in a situation like this and the building inspector then must retire to his firm-but-fair manner of operating. He simply must take the time, even at the expense of the project, to investigate the alternative being introduced so that he does not contribute to the degradation of the project.

A building inspector cannot be strictly a tradesman or strictly a design professional. He must have wide-ranging interests, and should have experience that will have exposed him to many different situations. He must be interested in continuing his education, learning about the new materials and systems, and keeping abreast of the changes in the building codes.

Most educational programs for building inspectors are continuing education opportunities. A few schools now offer full certificate or degree programs in building inspection, but the vast majority of inspectors are hired from the trades or professions. Basic initial instruction is required, with the other training sequences acting as booster or reinforcement devices. The building inspector should be exposed to and trained in:

1 Basic code usage and interpretation.
2 Periodic code updating.
3 Techniques of inspection for each major material system.
4 Public relations.
5 Written and oral communications.
6 Energy code provisions.
7 Relation to other codes of the jurisdiction.
8 Newly introduced material and construction systems.
9 Department or regional operating procedures.

Several efforts are being made for the formal certification of building officials and inspectors. The first system is in place in New Jersey, and recently some inspectors have lost their jobs because they were not able to achieve the certification required by the state. Other states are moving in this direction, while others have chosen to use a voluntary system. Both initial and recertification requires education sequences. The certificate must be renewed periodically, with continuing education being a requirement. Although demanding, this will place a great deal of emphasis on professionalism in the ranks of the inspection corps.

Basic to construction inspection is the principle that such inspection should be progressive, that is, in keeping with the progress of the job. Once an item has been inspected and found to comply with all regulations, it is in effect given approval and can be so recorded on the inspection records. Subsequent inspections need not include these approved items unless there has been some damage, addition, or alteration to it. It would be virtually impossible and a useless exercise to inspect every item on every inspection trip in addition to inspecting the new items just recently installed. It may well be that an item can be installed early in the construction and receive approval, only to have a finishing procedure, material, or item added later in the construction. A final approval, or at least another approval, would be necessary. However, between the initial approval and the next "inspectable" feature, there would be no inspection except as noted. The principle of inspecting every item each time a site is inspected would be costly, burdensome, unnecessary, and for the most part slow and counterproductive to the completion of the project.

As with the other inspection participants, the building inspector must keep pace with the project and must impartially and thoroughly inspect each project for which he is responsible. In other words, the inspector cannot allow certain projects to become his pets, receiving his easy approval or extra inspection time simply because he is intrigued by the construction of the project. He must keep pace with the progress of each job. He must make sure that all the necessary requirements of each job are met, and he must see to it that all the inspections are properly carried out. This must be done day in, day out for each project. Methods vary among departments, but it is essential that each project receive proper and timely attention.

It is a very imposing list of characteristics that the inspector should carry into his inspection techniques. Basically, he should be consistent, continuous, timely, well founded, well

Figure 41. A $20 million college library took 30 months' construction time, even with a construction manager and fast-tracking. This requires constant and diligent inspection by all.

documented, firm but fair, relevant and proper, thorough, persistent, courteous, and knowledgeable. It may seem that some sort of a super human boy-scout type is required to accomplish all this. An inspector who is interested in his work and who has confidence in his superiors and in the basic system itself, can easily accomplish all these aspects of inspection, but he must somehow have these characteristics as part of his nature.

An inspector who is very knowledgeable but at the same time very abrasive in his dealings with others can become quite a problem to the building official and the governing body of jurisdiction. This is not to say that every inspector should be an easygoing, happy-go-lucky fellow who may be short on knowledge; then nothing is gained because the basic system or proper construction is not being served. The inspector's demeanor and action should be proper for each condition he faces.

The inspector must be able to change as the situation at hand warrants. For the most part, projects will run very smoothly, and the contractors will prove very cooperative. However, some situations arise where a contractor feels that his rights are being denied or his method of operation is being challenged. Here the inspector must be able to draw upon every resource at his command to resolve the problem as quickly but as satisfactorily as possible. A well-founded (properly based) knowledgeable and soft-spoken approach will solve the vast majority of the problems, but the inspector should never be afraid to use other tactics if necessary. He can involve other personnel, supervisors, formal orders, the appeals process, and, of course, formal legal action. The wise inspector keeps these options in proper perspective and in proper priority so that he does not use "an elephant gun to solve a rabbit-sized problem."

THE AUTHORITATIVE BYSTANDER

The building inspector, since he does not participate directly in the design or the construction of the project, is a bystander to a large extent. To many people, this posture is a source of basic irritation. "Who is he to tell us how to do something?" they say. The intent and basic charge of inspection is not to tell others how to build the project, but to see that the applicable minimum standards are met. The inspector must know the code. Everyone involved in the project is well trained in his discipline, but little formal instruction is available regarding the principles, intent, and operation of the various codes.

Traditionally, code experience was gained through trial and error, work stoppage, and field changes. More recently design professionals have recognized the constraints of the code, and make a code search before beginning their project design. In this way, the code provisions can be more easily accommodated, and the design will not be disturbed by a late revision. Design professionals can help themselves by making early contact with the building department plan examiner to establish basic understandings and to receive the necessary interpretations of the code provisions.

This attitude may also carry over to the field inspector. Contact between the design professional and the inspector can be mutually beneficial. The field inspector can quickly carry any problems back to the office, have them discussed, and promptly present solutions to the design professional. This is certainly a much more positive and prudent approach than to try to deal with the codes as an afterthought or when a problem is discovered. No one in the inspection system, particularly the building inspector, is interested in posting a stop-work order on any project. It is as much in his interest to see that the job progresses as it is for anyone else. When he does not find the design professional or the contractor cooperative, he may have to resort to more drastic actions. The inspector must be able to analyze the situation and apply the proper personal and inspection techniques to resolve the problem. No one can predict the reaction of another; a design professional can often be as hard-nosed as the most aggressive contractor. Proper training and experience will stand the inspector in good stead under any circumstance.

The problem with some building departments is that either through the personalities and reputations of the people involved or through their efforts to reflect the attitude of their governing body, they become so restrictive, bureaucratic, and uptight with their position that they become obstructions not only to an individual project but to the overall development of their community. The reputation of a community is on the line each time a building inspector opens his mouth. If the reputation gets too bad, larger developments, expansion, and possibilities of increased tax base may well vanish because the new developers simply do not want to deal with the people in that jurisdiction. This creates a very drastic situation for everyone involved.

Although the building department may be seen as a bystander, it does bring a certain expertise to the project that would otherwise not be there. Therefore, it not only behooves the design professional to extend himself to the building official; the building official and all of his personnel should extend themselves to the design professional and the contractors. The spirit of cooperation that ensues can greatly enhance the entire development picture of a community, and, of course, it can greatly ease the construction of an individual project.

Professionalism within the ranks of the building inspection community has become a daily watchword. It is no longer considered good practice to allow an unqualified but well-liked individual who has been an active participant in the community, to serve out his few remaining working years as a building inspector for the county. This system saw building departments manned by people who were physically incapable of making inspections, who were inflexible and unknowledgeable and who, for the most part, collected their paychecks rather than becoming viable public employees.

The complexity of construction, the development of communities, the continuing hazard of one use to another, the introduction of new materials and systems, and the entire economic and social life of the community are changing so rapidly that only a professional can handle the situation. More and more systems of certification and continuing education for not only building inspectors, but also for code administrators, are coming to the fore. Basic criteria for hiring personnel are becoming more and more selective and restrictive. Simply put, a person must bring much more to the job than was required previously.

Continuing education is an absolute must. Remedial education, simply going over the existing policies and procedures of the department, must also be an ongoing process. There is still a tremendous need for early education in the principles of the code, such as exiting, fire prevention, fire protection, and the like. Once the basic provisions of the code are understood, they can easily be incorporated into an inventive design. Not only must the design professional be educated, but the building inspector must be as well, so both can see how problems can be solved.

Today, though, many dislike some provisions of the code and will insist on trying to cheat on these items in every design. An architect may simply detest stairs, and his designs always merit a very close look because his exit stairs are always suspect in one way or another. What has he basically gained or achieved? What is he really trying to do? Certainly, there is an acceptable feature in each design that can accommodate proper exit stairs. Besides, a beautifully designed project can be as disastrous as a poorly designed one if basic safety requirements are ignored or toyed with.

The design professional, like the building inspector, is always and intimately involved in public health, safety, and welfare. In many states, registration laws place this burden on the design professional. Such laws demand that the professional design and execute the entire project within code limits. Further, he is to report to "the building official" any acts he knows of or observes that are contrary to the code. Basically, the design professional is placed in the "public health, safety, and welfare" business, and he can't avoid this responsibility without violating the very law that allows him to practice. What is gained by trying to shortchange the occupants of a project, simply for the sake of a design? Nine times out of ten, any shortchanging will be caught, and the remedy for this noncompliance could destroy the design. However, chances are very excellent that with a little extra effort the design concept can be maintained while still adhering to the code.

Codes are not meant to impinge or inhibit design. Each code provision must be met, and it can be met in any one of numerous ways.

Here again the flexibility of the building inspector, the code administrator, and the design professional comes into play. The more interplay, the more understanding; and the better the relationship between the design professional and the building officials, the easier it will be to find a satisfactory solution for any design problem.

However, the inspection system, primarily the governmental inspection system still must exist. The presence of the building inspector on the job is mandatory and is still the major function of the inspection system. Recently building officials have seen an erosion in their immunity to liability (personal tort liability). There seems to be a movement away from governmental immunity, which will make the official walk a fine line. Each jurisdiction must establish its own position in this regard, but the inspection system should be ready to be evaluated according to the level of compliance, the timeliness of inspection, the quality of inspection, and the volume of work accomplished.

The entire process of construction in the United States has some counterproductive aspects. The bidding process, for instance, which demands that the low bidder be hired for the job, can play havoc with any project. The contractor can very well make an error in his bidding calculations. His normal tendency is to try to recover the money reflected in the error so that he will not lose money during the

Figure 42. Projects of several multi-unit buildings often want to be occupied as each unit is finished, thus a need for a change in standard inspection policy. Insure safety, and code compliance before any occupancy, in any event.

Figure 43. A "cheat." There is let-in bracing in one wall, but none in the other, and little chance for it with the window placement. Also, note the hole in sheathing, which is an energy code violation.

construction of the project. In trying to do this, he often will, in one way or another, adjust his methods or "cheat." As long as there are people who have this attitude, there will be a need for inspection at all levels.

In some instances, complying with the code requirements adds expense to the project. In trying to adjust his cost a contractor may try to cut out some of this added cost. Again, this points up the need for the governmental inspector to see that the basic minimum standards are achieved. If there must be cutting on the job, if there must be adjustments, it will have to be done in some area other than code compliance. Of course, the architect must police the contract to see that the construction documents are abided by and that the owner is given full value for the job, including the code requirements. In Europe the prevailing bidding process whereby the high and low bidder are eliminated and the award is made to the bidder who comes closest to the average of the remaining bids produces a contractor whose contract price is comfortable. The bid is one the contractor can live with, without fear of encountering gross errors in his estimate later. The bid is the right one for the project.

Surely this system produces fewer adjustments and less need for an inspection system or a policing of this contractor. He knows that by doing the work in abidance with the code and the other documents, the cost he has calculated will allow him to make a profit, barring any unforeseen problems on the job.

In the private sector, the owner often will have a private list of bidders—people recommended to him or people with whom he has dealt in previous projects, who can be relied on to give proper pricing and at the same time will produce a code-abiding project. This system tends to eliminate the "bad" contractor and gives the owner some assurance that his project will proceed and will be completed in good fashion. Although this system is not 100%, foolproof, the owner is more comfortable in this situation.

The building inspector, by not participating in the bidding process, many times will not know the intimate details of the cost of the project. He will, through his experience, find good contractors and bad contractors—people who are cooperative and produce good work without any coercion, as opposed to those who are always looking for an "out." He must adjust his inspection technique to the character of the contractor. If he is known to be a devious, cheating person, the inspector will have to keep a closer watch on the project. He will have to stand firm in enforcing any orders that are given to this contractor.

Figure 44. Plywood used for corner bracing, with better insulator (and cheaper) fiber board used for infill. This good solution keeps all things in mind.

Faced with a very bad situation, the inspector must be fully prepared to follow the proper channels in resolving the problem. If a mild approach proves fruitless, he must contemplate issuing formal instructive orders, stop-work orders, or using appeals, injunctions, or other court actions. He must involve his supervisors in the proceedings and seek the advice of the local legal counsel as necessary. A good department will train its personnel in the proper court procedures and the proper method of testifying in court. The inspector should issue proper oral and written orders and have a good job history to introduce in court. All this is time-consuming and often frustrating, but the inspector must remember his basic charge and must be ready to do whatever is necessary to carry it out in a precise and professional manner.

All of this becomes part of the job of the inspector. The building inspector should allow nothing to separate him from his basic charge—the protection of the public interest. He must understand that he is doing something the general public cannot do for itself. That's why he is hired; that's why he is paid by tax money.

The building inspector's function is not all one-sided. There is a possibility that the building department can, in fact, also be of benefit to the contractor or the developer. If someone is injured or dies while on the premises, or if there is some sort of litigation, the records of the building department may be subpoenaed by the court. Surely, an owner, developer, or contractor who can produce documentation that shows that he followed proper procedures and that his project is code-abiding (that he did, indeed, take the public interest into account), will find that those records may well stand him in good stead with the court. A person who has a record of constant code violations, constant conflict with the inspection system, or shutdown of his job because of deviousness and poor construction is not going to have the same good reputation to present to the court. Therefore, one aspect of building inspection that cannot be denied and that can never be underestimated is recordkeeping. Every project that comes into the building department and is under its control should have a thorough history: a record of all reviews, and approvals; a record of any variations or revisions; a record of all inspections; a record of any orders and any extenuating circumstances; and a complete chronological history of the project from start to finish. As noted before it is usually required by statute that the building department keep the history of the documents for a number of years. They are public documents and are open for review, and they can be used on either side in court proceedings.

To be of value, the inspection history should be well founded, free of guesswork, off-hand comments, opinions, and irrelevant information. It is simply not enough to record that on a certain day the inspector was, indeed, on the job and looked at a particular portion of the construction. Behind his initials and the date of his inspection record should be a checklist enumerating all the items that were checked and found to be in compliance before his approval was initialed on the record. The inspection system cannot be arbitrary, imposing more on one job than on another. Each project must have the same items checked in the same manner as any other project. Flexibility must be built in for items that do change from project to project, but there are basic construction elements that are the same or very similar on every project. In this manner the opinion of the building inspector is taken out of the picture. He has no choice in what he is going to check; his choice comes down to how does this comply and does it comply.

Figure 45. Does the code permit this solar house? How must the inspection change to accommodate this new system.

Of course, it's possible for fraudulent documents to be filed. Over a period of time, the type of person who would engage in such activity is going to be found out, and he should be summarily dismissed. An inspector who has lost sight of his charge, who has simply chosen to "windshield" inspections (from his car), does not want to get his clothes dirty, does not want to "get up" into the project, and does not want to follow the basic checklist and the operating procedures of his department is of no value to the department. He simply is not doing the job.

There may be other problems with building department personnel. A plan examiner who is extremely lenient with interpretations of the code is just as bad as a field inspector who closes his eyes to substandard construction. The basic charge of the department—the protection of the public health, safety, and welfare —must be met. Leniency fosters disaster; further advantage will be taken of lenient personnel. The public cannot enforce codes for itself; the government hires the building inspectors to do that work.

It is baffling that some people involved with construction try to build projects that don't

68 Systematic Construction Inspection

meet the code restrictions. Everyone has the attitude that "it can't happen to me." But it does! We have learned little from the tremendous building disasters in our history. We still build to meet the law (in most cases) but forego for the sake of convenience, the protection it offers. Despite codes and inspection that are better than those of the past, lives are still being lost due to causes that would be silly if they weren't so tragic. There must be constant improvement in attitudes and an intensification of interest among the building department staff, and the staff's education along job lines must be upgraded.

Figure 46. A new shopping center, which presents a unique inspection problem. The inspection sequence is protracted over a long time until all occupants are in place. Protection between occupancies must be ensured, even as the construction continues.

No building department should take on a pious, know-it-all attitude, assuming that contractors or others are wrong until they prove themselves right. The staff is hired by the citizenry; its charge is to protect the citizenry. The building department must be acutely aware of this in its handling of all inspections and all other departmental functions. The inspection procedure can become very sensitive because it regulates homeowners in their "castles." It forces them to take action to meet the code, where they feel it is not necessary. It makes them spend money for additional or remedial work or for additional inspections. Inspection is as much a job of public relations and public education as anything else.

Perhaps one of the greatest problems to any building inspection department is the lack of an educated public. It would be very helpful if every student during his required years of schooling somehow was given some basic instruction about the laws of his community. It certainly seems important that students be made aware of their tax situation as well as the pricing of various commodities, cost evaluation for the consumer, and the basic laws by which they live. The schools best serve students by instructing them in how to protect themselves a little more or making them more aware of the laws and the protection the laws provide.

Every student is well aware of the speed limit in his area and of the law that requires schooling to a certain age; but we could build a better society if we could also make students aware, in a general manner, of more of the laws affecting their daily lives. They would become more aware citizens and might demand more of their laws, elected officials and enforcement agencies. Further, they might support these efforts in a larger measure.

Until the public is taught about the laws that affect them, an inspection system, primarily the governmental building inspection system, is an essential part in all construction. Only through public education and public awareness will meaningful changes be made in public support and funding for proper inspection activities, which can directly protect the citizenry from hazards in all structures. They must demand excellence from the inspection program and demand a system of severe penalties for code violations.

CHAPTER 8
CONCLUSION

> You, Building Officials, carry with you a public trust of the highest order. You share, along with police and fire officials, a number of important duties. You enter the private homes, and the private lives of many of our citizens. Fire and police personnel usually arrive on the scene *after* something has gone awry. You get on the scene in an attempt to head off and avert trouble and disaster. It is for this reason that you people serve on a higher plane than most public servants. To most of our citizens, you are the only non-emergency personnel in public service they ever meet. You deal in more than service. You provide the understanding and the expertise which most of the citizens do not possess, but which each sorely needs. You are servants of the public of the highest calling. You are the couriers, and carriers of the public trust, and being that, you are the synthesis of government in action, and the epitome of our democratic processes, personified.
>
> Edward C. Vavreck
> Assistant City Attorney
> Legal Counsel, Department of Inspections
> Minneapolis, Minnesota

Despite the constant and voluminous outpouring of sound, sophisticated technology for the construction industry, the United States continues to see an intolerable number of lives lost due to fire and other building-related incidents. The technology is not failing, people are. Recent figures indicate that almost 50% of all fires are directly attributable to human error, but these errors account for 75% of the monetary loss due to fire. (In 1979, this amounted to $5.75 billion.) We are our own worst enemy. Years may pass without major disasters, but they still occur all too frequently. This is repugnant to a society that places great value on the individual life.

We regret the loss of each of the 9000 lives lost each year in this needless fashion, yet we continue to see a tragic parade of fires that result in multiple deaths. We continue to live, shop, congregate, recouperate, be entertained in, and encourage others to use facilities that can be patently unsafe. Just as crossing the street is an everyday gamble, so is entering any building. The sooner the general public is aware of this, the sooner we have a chance to reduce fire loss. Recognition should trigger more and continual calls for help, safety, and compliance. Hazards develop or have been created, and for the most part we are unaware of them. Many would not recognize hazards if, indeed, they were to confront one, since the general public is not well educated in safety and codes in general. After each major tragedy, the public is outraged and then sensitized; they become aware of exits, look at wall finishes, speculate on construction, warn their children; but all too soon we go back to our normal routines, going about our business unconcerned about the buildings we live and work in (taking the attitude that what we can't see or don't recognize won't hurt us). The public simply must be educated about safety so that

Date	Structure	Place	Deaths
November 28, 1942	Night club	Boston, MA	492
June 5, 1946	Hotel	Chicago, IL	61
December 7, 1948	Hotel	Atlanta, GA	119
December 1, 1958	School	Chicago, IL	93
December 8, 1961	Hospital	Hartford, CT	16
November 23, 1963	Nursing home	Fitchville, OH	63
December 23, 1963	Hotel	Jacksonville, FL	22
February 7, 1967	Restaurant	Montgomery, AL	25
April 5, 1967	Dormitory	Cayugo Hts, NY	9
February 25, 1969	Office building	New York, NY	11
January 9, 1970	Convalescent home	Marietta, OH	31
March 20, 1970	Hotel	Seattle, WA	19
September 13, 1970	Hotel	Los Angeles, CA	19
December 20, 1970	Hotel	Tucson, AZ	28
January 14, 1971	Elderly housing	Buechel, KY	9
October 19, 1971	Nursing home	Honesdale, PA	15
January 16, 1972	Hotel	Tyrone, PA	12
November 30, 1972	High rise building	Atlanta, GA	10
June 24, 1973	Cocktail lounge	New Orleans, LA	32
September 13, 1973	Nursing home	Philadelphia, PA	11
November 15, 1973	Apartment building	Los Angeles, CA	25
June 30, 1974	Discotheque	Port Chester, NY	24
June 9, 1975	County jail	Sanford, FL	11
December 18, 1975	Night club	New York, NY	7
January 10, 1976	Hotel	Fremont, NB	20
January 30, 1976	Nursing home	Chicago, IL	24
February 4, 1976	Apartment building	New York, NY	11
August 12, 1976	Oil refinery	Chalmette, LA	12
October 24, 1976	Social club	Bronx, NY	25
December 20, 1976	Apartment building	Los Angeles, CA	10
December 22, 1976	Department store	Brooklyn, NY	12
December 23, 1976	Apartment building	Chicago, IL	12
January 28, 1977	Hotel	Breckinridge, MN	17
May 28, 1977	Supper club	Southgate, KY	165
June 26, 1977	Prison	Columbia, TN	42
October 24, 1977	Cinema	Washington, DC	9
December 10, 1977	Hotel	Bay City, MI	10
December 13, 1977	Dormitory	Providence, RI	10
December 22, 1977	Grain elevator	Westwego, LA	36
December 27, 1977	Grain elevator	Galveston, TX	18
January 28, 1978	Hotel	Kansas City, MO	20
November 5, 1978	Hotel	Honesdale, PA	12
November 5, 1978	Department store	Des Moines, IA	10
November 26, 1978	Motel	Greece, NY	10
December 7, 1978	Tenement	Newark, NJ	12
December 29, 1978	Mental institution	Ellisville, MS	15
January 20, 1979	Apartment building	Hoboken, NJ	21
April 1, 1979	Boarding house	Connellsville, PA	10
April 2, 1979	Boarding house	Farmington, MO	25
April 11, 1979	Boarding house	Washington, DC	10
July 31, 1979	Motel	Cambridge, OH	10
November 11, 1979	Boarding house	Pioneer, OH	14
December 27, 1979	Jail	Lancaster, SC	11
July 26, 1980	Motel	Bradley Beach, NJ	23
November 21, 1980	Hotel	Las Vegas, NV	84
December 4, 1980	Motel	Westchester County, NY	26
February 11, 1981	Hotel	Las Vegas, NV	8

Figure 47. Partial Listing of Major Multi-Death Fires in U.S. Structures.

it will feel safe or uneasy, as the case may be, and will react accordingly. But how can we achieve this "feeling" in a mass of individuals with different sensitivities?

No one advocates that people be fearful whenever they walk into a building and that they feel safe only in their own homes. But we must seek and demand proper protection. Why don't we support fire department levies? Why don't we try to understand the inspection system when it touches us, and even demand more of it? Arson is a crime, but maybe we should make fire a crime; perhaps code violations should be crimes rather than minor offenses. Why don't we get strong court penalties, especially since only the worst cases are getting into the courts anyway? We don't seek these changes because not enough people are touched by the disasters and there is no organized effort, by a large segment of the population, to effectively "demand" change. The groups involved are relatively small and have very little influence; the "average" person and family is involved. Even a recent incident where a fire claimed the lives of several top executives from one corporation did not effect a massive change, although it may "trigger" some procedural studies by other corporations in an effort to prevent other such calamities. These small, isolated actions, however, may help in the long run.

Lives are lost every day in fires started by unknowing children playing with matches or candles. But lives are also lost, and at a more rapid rate, because of fires promulgated by code violations: holes in fire walls, flammable finishes, material producing toxic fumes, and locked exits or, worse, inadequate exits. We recycle and refurbish older buildings for new uses, and in the process old hazards may be glossed over, or covered by new materials or may have additional loads imposed on them. Although the hazards are unseen, they still exist, and eventually they may erupt and claim some lives. In all these situations unthinking humans are to blame.

Although the profusion of single-death fires is still repulsive, two major American disasters within the last few years vividly point out the need for the imposition of proper construction inspection on an apathetic public in occupancies other than residences (primarily in mass lodging and places of assembly).

The 1977 fire at the Beverly Hills Supper club, in Southgate, Kentucky, which took 165 lives, seems to point up the problematic situation of the owners' interests apparently being allowed to take priority over the interests of the using public. This facility, open to the public, was constructed after being designed by a professional. However, the documents were incomplete (not intended for construction), and the construction was not supervised by the design professional; nor were the permit system and the governmental inspection agencies properly involved. A grand jury investigation (reviewed by a special prosecutor) produced no indictments and, indeed, vindicated the owners. There was no conspiracy to violate the codes, it was found, and apparently the owners had done all they could to comply with the permit and review sequence. The jury, however, scored the inspection system of the governmental agencies (both local and state) for lack of understanding, communication, action, and initiative in following the construction. Basically, no one could be held responsible for the fire; it was just something that "slipped through a crack."

The owners built what they needed when they needed it, with the assumption that all work was code-abiding, under permit, and inspected. However, the result was a terrible disaster. Investigation points to the possibility of improper and insufficient exits; highly flammable interior finishes; toxic smoke production; inadequate fire stopping; little if any fire suppression systems, fire fighting equipment, or alarm system; improper wiring; concealed voids between segments of construction (with combustibles present); and overcrowding of the entire facility. Any one of these discrepancies could foster trouble, but in combination they caused disaster. Elegance, convenience, and mere accommodation replaced the top priority items of proper code-abiding construction and basic safety systems.

Who cares if the new carpeting is flammable and gives off toxic fumes? Nobody cares if the ten people in the waiting areas are seated, even though the room is crowded. Just put a few more chairs in the aisle (the only open space left) so these people can see the show (besides, they usually spend a lot). Nobody, it seems, was dissatisfied—until it was too late.

Not only had the inspection system obviously failed, it was never properly triggered. This was a massive facility in a very small town, which had no real handle on the hazard in its midst. The state jurisdiction was located some distance away, and most of the construction was carried on within the old walls. Given the confusion among inspection agencies, the lack of distinct jurisdiction, and the antiquated code that was the only tool available, it was amazing that the disaster was not of even greater proportion. A valiant effort by the town's fire department and those of many surrounding communities was the real "saving" grace. A far better and easier alternative to fire suppression and rescue lies with initial and continued compliance, inspection, and upgrading, along with owner and public consciousness of safety.

Inspection of existing facilities is extremely superficial at best, but the more obvious deficiencies could have been observed and orders issued. The basic construction, though, is hard to evaluate, especially when it is continual and old construction is covered over.

Overcrowding is a big problem in every jurisdiction, but to convict and penalize those responsible is difficult. The law and courts simply are not strict enough in these situations. If a liquor permit can be denied because of obscene performances, how about denial because of failure to meet and observe codes or for violating posted occupancy limits? When violations are allowed to stand, basically unchecked, lives can be lost.

The 1980 fire at the MGM Grand Hotel in Las Vegas, Nevada, which took 84 lives, was a major disaster, but with a different twist. The building, built in a high-powered, extremely active market, well protected by codes, was built to meet the law in effect in 1972 when construction began. How then could this facility be the site of a disasterous fire? Major changes in thinking have produced entirely new concepts of fire rated construction, fire suppression systems, fire alarms, evacuation of buildings, compartmentalization, refuge areas, and the like, within the last eight years. But the MGM Grand was behind the time; it met an old code. There was no provision, however, that demanded that a code-abiding structure be constantly upgraded to meet the latest code provisions. Simply, the MGM Grand could not be built today as it was in 1972.

Oddly enough, though, the cause of the disaster was again human in nature. Investigation points to code violations in the building that allowed the fire to spread and become a killer. Apparently, holes in fire walls, flammable finishes and furnishings, limited use of sprinklers, and lack of proper venting contributed to the disaster. Although the building met the code of its day, it was beyond the technology of that day, and the violations allowed a marginal situation to become bad.

Even with a thorough inspection system, few jurisdictions have the "clout" to demand that all buildings keep in line with the kaleidoscopic code provisions. This process would be endless, and prohibitively expensive. Who could afford to own a building that was constantly "under construction," installing the latest equipment? Yet doesn't the public health, safety, and welfare warrant this? At best, initial construction must comply, and periodic inspections (by any agency) are necessary to ensure continued compliance.

Some interesting facts have surfaced. The loss of life at Beverly Hills was 5% of the people in the building, whereas at MGM Grand it was only 1%. Perhaps the observance of the code is the factor directly involved. It surely seems that the effort to comply is worth it. In the last eight years, little more than a handful of fires (on average, eight per year) accounts for an average of 164 lives lost each year. But these are just some of the 249 multi-death fires that take 1,106 victims annually. This carnage is a minuscule 0.009% of the total fires (3,070,600 in 1978), but accounts for an alarming 13% of the yearly 8780 fire deaths. One can easily see that the number of fires is outragous and can understand (to some degree) death in ones and twos in many unregulated units, but there is no understanding of or excuse for the high death rate in highly regulated occupancies. Needless to say, the compliance and inspection systems are failing.

Construction and construction inspection are both imperfect sciences. The perfectly constructed building, with all elements being code-abiding, may never be built. As long as unthinking individuals see or feel the need to take shortcuts, there will be built-in imperfections. These imperfections sometimes allow problems to start, sometimes allow problems to continue and worsen and perhaps result in a disaster. The "innocent" scrapping of fireproofing from a steel column to allow a tighter fit of another material is an imperfection that led to disaster. The building caught fire and the other material burned away, exposing the column to the flames. The heat twisted the column, which was on the thirty-second floor of the building. In an instant, there is a major problem resulting from a small act.

Whatever measures are necessary must be employed by the governmental agencies and their inspectors to ensure compliance. Fees should be adjusted to provide adequate funding for a professional staff, proper departmental operations, and a diligent, complete inspection sequence. The penalties should be harsh, and the judiciary should recognize the hazard of light "wrist-slapping" as a symptom of disaster. A person in violation should expect to be caught and fined. Don't most people observe the speed limit for these reasons?

Much research has been done over the years, but still there are differences of opinion on just how the codes should be written and what techniques should be used to protect the public. There is, currently, no clear path that will result in maximum protection and still allow for modern construction techniques. This is not to say that all new construction methods and materials are faulty, but society has not yet solved the multifaceted problems of construction, energy saving, fire prevention, life saving, and economics. We seem to solve one problem, only to create others. Our research must be broadened, preferably under the sponsorship of one agency.

Building and fire codes should be set forth in such a positive and creditable form that any person who is involved with them can easily and clearly see his responsibility and the consequence of noncompliance. Ideally, voluntary enforcement (and compliance) is the result being sought through the coding and inspection program. Great, though, are the pressures to do otherwise. Unfortunately!

The inspection system is locked into the current code provisions, which, of course, tend to reflect the latest thinking and technology. Some time lag is normal between introduction of new systems and materials, and code acceptance, but everyone is party to seeking approval if it is to their benefit. Simply, the code must be enforced as it stands. In some instances owners will advance their protection on the advice of their insurance experts. This is no particular stroke of genius; it is amazing that every business venture does not seek maximum protection at least through compliance if not through more extreme measures. This would minimize liability, especially where the public is directly involved. Remember, the inspection system only sees that the current minimum provisions are not undercut.

Interesting situations sometimes occur. Currently, two new "clubs" are being constructed within a few miles of each other, not far from the ill-fated Beverly Hills. One is strictly a supper club, and the other is a club and discotheque. The supper club owner is refurbishing an old building and has maintained a high level of public relations about meeting and, indeed, exceeding the required code provisions (for instance, sprinkler heads 6 feet apart throughout). All furniture carries a one-hour fire rating. Small placards are being placed on each table with instructions on fire reporting and evacuation, as well as with the following information, which reads:

For your safety, this building is constructed with:
 1. The most modern fire and safety devices.
 2. Complete sprinkler system.
 3. Adequate fire exits and handicap ramps.
 4. All employees trained to assist in the event of an emergency.
 5. An alarm system connected to fire and police departments.
 6. All doors closing when an alarm is pulled, but they will not lock so you may exit if necessary. This is to contain a fire at its origin.
 7. All doors and walls fire rated 1-hour.
 8. Furniture and furnishings fire resistant and nontoxic.

The club/discotheque is being built by the owners of the Beverly Hills club. They are building within the corporate limits of the City of Cincinnati (within a few blocks of City Hall and the central fire station as some have pointed out). They are meeting every code provision and every requirement of a fully certified building department. They have seen fit to use nonunion labor in an area where the union workers perform the overwhelming amount of commercial construction work. This is being held against them as another example of how they are "cheating" and allowing substandard work (the innuendo is "hazardous") These are baseless allegations, and it is hoped that time will not prove the reverse.

74 Systematic Construction Inspection

It will be interesting to follow these two construction projects and how they are accepted and supported by the public. Despite in-depth investigation of the Beverly Hills fire, the more recent tragedies have resensitized the public, and this may work against the new venture. However, it is worth watching the supper club venture to see how long the present consciousness of safety is maintained, how long the table placards are used, and just how successful the operation becomes.

Many times the building as constructed will meet the code, but the use of the building (or rather misuse) will produce massive adverse results as happened at MGM Grand. Narrowness of mind and trying to solve the immediate need at the earliest time, in the cheapest way, cause people to negate the best fire safety features and to add substantially to the fire loading of the building. Building codes have no method to prevent this, and fire codes are just now emerging that try to handle these situations. Obviously, stiffer regulation is at hand; it is hoped that the full support of the public and the courts is forthcoming as well.

In a 1977 instance the stockpiling of material in a West German warehouse to extreme heights canceled out the effectiveness of the sprinkler system. The result was a massive loss ($100 million to building and stored materials) because the water from the sprinklers could not reach the fire down in the stored materials. Human error, again and the need to take a shortcut to solve a problem are responsible. The building, though, met the codes.

Oddly enough the building occupancy and use are the basic hazards to a building and its occupants. A structure built to be 100% safe would be highly undesirable (one could not live comfortably in it) and would demand constant supervision of the occupants by an unyielding, uncompromising authority. Bad habits and any tendency toward bad housekeeping, cluttering, and taking shortcuts are the real hazards. These are not controlled by building codes, although some aspects are now being controlled more and more by fire codes.

Since it is well established that owners and occupants are going to "abuse" buildings, it is imperative that the inspection system demand absolute adherence to approved documents. Chances are the building will never be safer and more code abiding than at the moment just prior to occupancy. In our current economic climate, maintenance and upgrading are relegated to very low priorities, while convenience, attractiveness, and money-saving ventures are pushed at every turn.

Figure 48. A safe building. Construction is finished, and the inspection sequence is completed. The photograph was taken the day before occupancy.

Fires continue to claim lives. Youngsters will still take the batteries from the smoke detectors so they can furtively smoke a cigarette; some people will smoke in bed, and equipment will be poorly maintained. Sick minds will continue to set fires, and antiquated and deteriorated materials will erupt in flames. But we must try to achieve the best possible solutions, even though the deck may be stacked against us.

The tremendous pressure of time often forces (or tends to force) acceptance of conditions that are changed or that are not in compliance, particularly with regard to material substitution or systems deletions because "we can't get it"; "it's in transit and lost"; "they're on strike"; "its not available here"; "out of stock"; "it has to be a special run"; "this is just as good as"; "thought we had it in our warehouse"; "can't afford to wait for the right stuff"; "who will know the difference?"; "can't stop now"; "it'll cost me a bundle." These are real problems, yes, but hardly grounds for not following code provisions.

The pressure of the economy on the contractor and developer is great indeed. No job can afford a slow-down or suspension because of the intolerably high interest rates on construction loans. Poor planning and unrealistic scheduling, which promises project delivery without consideration for lost time, tend to foster shortcuts, illegal substitutions, and shoddy, noncomplying work, but these are not viable alternatives to proper code-abiding construction.

One of the best instruments at the command of the inspection system is the Certificate of

mandated regulations are not needed and would only confuse the code situation, and provide unenforceable provisions, since there is no national building department. Local control should remain, but the local codes must be reviewed, updated, and strengthened, and *fully enforced.* The same recommendations are made over and over:

1 Make thorough inspections during and after construction.

1.04 CONTRACTOR'S CLOSEOUT SUBMITTALS TO ARCHITECT:

    a.   Submit evidence of compliance with requirements of governing authorities:

          1.   Certificate of Occupancy.

          2.   Certificates of Inspection:

                (a) Freight elevator.

                (b) Electrical system inspection.

                (c) Plumbing system inspection.

                (d) Sprinkler system inspection.

Figure 49. Excerpt from typical specification. A listing of the terminal documents required ensures that all work is properly touched by the inspection system, right up to initial occupancy.

Occupancy. This document can be withheld (with proper legislation) until all approval agencies have inspected the work and deem it safe for use. This program must be strongly enforced, and both occupant and contractor held liable for improper occupancy, thus ensuring that the public receives the best effort of the inspection system.

Many things must happen, and happen soon. There is a basic and vital need for cooperation and understanding. Sniping gets nowhere; laments about the past produce nothing. We need open discussion among the design professional, the contractor, and the code official, as this should foster even better cooperation and coordination.

Although concern about building safety is reaching federal levels (reflected in Congressional investigations and hearings), nationally

2 Enforce posted occupancy limits and impose strong penalties for violations.
3 Install direct alarm systems to fire departments.
4 Adopt revised national model codes in a timely manner.
5 Retrofit old buildings with safety systems (now or when remodeled).
6 Upgrade inspection system with more professionalism, proper budgets, and other support.
7 Make unannounced inspections during peak and off-hours of buildings used for public assembly.
8 Obtain personal liability protection for inspectors.
9 Permit only minimal and sparing grace periods for compliance.

The public must be educated about safety. Some building departments are using public service television announcements to let the public know what is required in the way of permits and inspections, and how the services can be received. Manufacturers' associations are producing more and more literature explaining new research, new products, and new uses for old products. More and more building officials' groups are forming active community groups. We must tap every resource at our command.

Although budgets are being restricted and cut, the charge remains unchanged and perhaps is more urgent now than ever. We must do for the public what it can't do for itself. The inspection system must demand the most from every element of the imperfect construction process; it must be flexible and accommodating, yet ever unyielding and diligent. *There is no other way!*

APPENDIX

The examples in this appendix combine to form a set of documents used by one building department. They illustrate the material involved in compiling a proper job history for each project for which a building permit is issued. The material is unique to the department, not because the forms and lists are original, but because the data collected is satisfactory to the jurisdiction and has proved adequate in various enforcement actions. Each jurisdiction must ascertain for itself what is essential, required and valuable for departmental operations and records.

DEPARTMENT OF THE BUILDING COMMISSIONER
HAMILTON COUNTY, OHIO
312 TEMPLE BAR BLDG., 138 EAST COURT STREET
CINCINNATI, OHIO 45202
PHONE: 632-8362

WHAT DO YOU NEED
TO APPLY FOR A
BUILDING PERMIT

APPLICATIONS

1, 2 & 3 FAMILY RESIDENCES
Fill out each form with a pen
- White Form No. 1 (Application for Permit)
- Heating Application Form "M"
- Energy Conservation Forms "E" and "E-6"
- Certificate of Occupancy (Application for Final Inspection)

ALL OTHER CONSTRUCTION
The same forms listed for 1, 2 & 3 family are required. Bring four (4) extra prints of drawing showing sanitary and storm drainage. Bring one (1) set of drawings for each of I.B.I. and Fire Department.

DRAWINGS

Working drawings sufficient to explain the extent of the construction, the materials used and how the building functions must be submitted. Normal construction would include:

SURVEY - Prepared by a Surveyor registered in the State of Ohio. The survey must explain the property, its physical properties and location, the location and size of the proposed structures, and locate and describe any paved areas leading to the structures.

FOUNDATION & FOOTING PLAN - FLOOR PLAN - Showing each floor layout, exits and windows, equipment location and describing the use of each room.

EXTERIOR ELEVATIONS - Showing all sides, including the footing and grade lines.

CROSS SECTION - Drawn thru the entire building, or wall sections. Sufficient detail must be given to understand the structural parts of the building and connections and fastenings.

DETAILS - Fireplaces & stairs require section thru them explaining their construction. Show insulation, sealants, weatherstripping, etc. and "U" factors.

SPECIFICATIONS - A separate listing of materials, their quality and method of construction may be used to further explain the items shown on the plans & sections.

** NOTE: The drawings and specifications must contain the information necessary to comply with the code requirements. The use of "LEGAL" or "PER THE CODE" is not accepted as being in compliance.

Drawings must be submitted in the form of prints. No marking on prints will be permitted.

Drawings must be at least 1/8" = 1'-0" (except the survey, which must be at least 1" = 50'-0").

The name of the owner and the person who prepared the drawing must appear on every sheet.

DEPARTMENT OF THE BUILDING COMMISSIONER
HAMILTON COUNTY, OHIO

312 TEMPLE BAR BLDG. • PHONE 632-8362
138 EAST COURT STREET • CINCINNATI, OHIO 45202

OWNER _____

CONTRACTOR _____

JOB LOCATION- _____ TOWNSHIP _____

APPLICATION NO.

FOR REVISION OR FURTHER INFORMATION REGARDING THIS APPLICATION, YOU MUST REFER TO APPLICATION NUMBER

CHECK ITEMS ATTACHED-OR NOTE AT RIGHT ⟶ REMARKS

- ☐ CONSTRUCTION DRAWINGS
  - ☐ 1-2 FAM. RES. ☐ OBBC
- ☐ LOT SURVEY
- ☐ BUILDING PERMIT APPLICATION
- ☐ MECHANICAL PERMIT APPLICATION
- ☐ ENERGY CODE SUBMITTAL
- ☐ CERTIFICATE OF OCCUPANCY APPLICATION
- ☐ SWIMMING POOL PERMIT APPLICATION
- ☐ FENCE OR SIGN PERMIT APPLICATION

IF ITEMS BELOW ARE REQUIRED AND NOT SUBMITTED WITH THIS APPLICATION, APPLICANT MUST UNDERSTAND THAT PERMIT WILL NOT BE ISSUED, AND WORK WILL NOT BE ALLOWED TO START UNTIL REQUIRED ITEMS ARE SUBMITTED AND APPROVED.

☐ PLUMBING RELEASE-FORM	☐ CONTACT PLUMBING DEPT., BOARD OF HEALTH 1016 TEMPLE BAR BLDG.	☐ APPLIED FOR
☐ DRIVEWAY APPROACH PERMIT	☐ CONTACT PERMIT DESK, COUNTY ENGR. 800 TEMPLE BAR BLDG.	☐ APPLIED FOR
☐ ZONING CERTIFICATE	☐ CONTACT TOWNSHIP ZONING OFF.	☐ APPLIED FOR
☐ ELECTRICAL RELEASE	☐ CONTACT INSPECTION BUREAU INC. RM. 503-C, 222 E. CENTRAL PARKWAY	☐ APPLIED FOR

APPLICATION RECEIVED BY: _____

NOTES:

CARD SENT: —

PHONED: —

FEE

KEEP THIS RECEIPT FOR YOUR RECORDS

HAMILTON COUNTY, OHIO
DEPARTMENT OF THE BUILDING COMMISSIONER

PERMIT APPLICATION FOR:
NEW BUILDING, ADDITION, ALTERATION
REPAIR, WRECKING, MOVING.

Initials of Cashier

CASH	CK.	PLANNING APPROVAL	ZONE	BOOK	PAGE

APPLICATION NO.

DO NOT WRITE IN THIS SPACE

APPLICANT – Complete all applicable spaces on this form side **USE BALL POINT PEN OR TYPEWRITER**

(DESCRIBED AS)
1. (a). N S E W side of _____ , _____ Street-Zip Code _____ St. No. _____
   (b). _____ feet, N S E W, from intersection of _____ Lot Number _____ Par _____

2. New Subdivision only _____ Township _____ Section No. or Municipality _____

3. **IDENTIFICATION**

	NAME	STREET ADDRESS	CITY	STATE	ZIP CODE	PHONE NO.
OWNER						
CONTRACTOR						
PLANS BY						

ALL APPLICANTS COMPLETE A THROUGH D

A. TYPE OF IMPROVEMENT
1. ☐ New Building
2. ☐ Addition — Enter number of dwelling units
3. ☐ Alteration — Enter number of dwelling units added _____, or deducted _____
4. ☐ Repair, replacement
5. ☐ Wrecking (demolish); enter number of dwelling units deducted _____
6. ☐ Moving
7. ☐ Other _____
   (SPECIFY)

Describe briefly proposed work: _____
_____
_____
_____

B. OWNERSHIP
8. ☐ Private
9. ☐ Public (Federal, State, Local)

C. COST (Omit Cents)
10. Estimated cost of improvement for which this application is being made: $_____.00

D. TYPE OF USE

RESIDENTIAL
11. ☐ One family
12. ☐ Two family
13. ☐ Three family
14. ☐ Four or more family Enter number of units _____
15. ☐ Transient Hotel or Motel Enter Number of units _____
16. ☐ Accessory Garage
    ☐ Car Port ☐ Tool Shed
17. ☐ Swimming Pool ☐ Above Gr. ☐ In Gr.
18. ☐ Other _____

NON-RESIDENTIAL
19. ☐ Amusement, recreation, place of assembly
20. ☐ Church, other religious building
21. ☐ Industrial, storage building
22. ☐ Parking Garage
23. ☐ Accessory Garage
    ☐ Car Port ☐ Tool Shed
24. ☐ Service Station, repair garage
25. ☐ Hospital, institution, nursing home
26. ☐ Office, bank, professional
27. ☐ Public works, utility building
28. ☐ School, college, other educational
29. ☐ Store, other mercantile, restaurant
30. ☐ Swimming Pool
31. ☐ Tank, tower, sign structure
32. ☐ Other _____

33. State in detail all existing and proposed uses of this building and premises:
☐ Existing _____
☐ Proposed _____

COMPLETE ALL ITEMS IN E FOR NEW BUILDINGS AND ADDITIONS ONLY

E. PRINCIPAL TYPE OF FRAME
34. ☐ Masonry (wall bearing)
35. ☐ Structural Steel
36. ☐ Wood Frame
37. ☐ Masonry Veneer
38. ☐ Reinforced concrete
39. ☐ Other

AIR CONDITIONING
40. Will there be Air Conditioning in this building?
    ☐ Yes ☐ No ☐ New ☐ Existing

TYPE OF HEATING FUEL
41. ☐ Electricity
42. ☐ Gas
43. ☐ Oil
44. ☐ Coal
45. ☐ L.P. Gas
46. ☐ Other

TYPE OF SEWAGE DISPOSAL
47. ☐ Public Sewer
48. ☐ Private system (septic tank, etc.)

TYPE OF WATER SUPPLY
49. ☐ Public
50. ☐ Private (Well, cistern)

FOR RESIDENTIAL BUILDINGS ONLY
51. ☐ Number of bedrooms _____
52. ☐ Number of bathrooms _____
53. ☐ No. of off-street parking spaces _____

FOR NON-RESIDENTIAL BUILDINGS ONLY
54. ☐ Number of off-street parking spaces
    (a) Enclosed _____ (b) Outdoors _____

The owner of this building and undersigned, do hereby covenant and agree to comply with all the laws of the State of Ohio and the resolutions of the County of Hamilton, pertaining to building and buildings, and to construct the proposed building or structure or make the proposed change or alteration in accordance with the plans and specifications submitted herewith, and certify that the information and statements given on this application, drawings and specifications are to the best of their knowledge, true and correct.

Application by _____ Address _____

DO NOT WRITE BELOW THIS LINE (Office Use)

55. Number of stories _____ 56. Total Sq. Ft. Area _____ 57. Total volume (cubic feet) _____
58. Zoning Approved by: _____ Date _____
59. Construction Dwgs. Approved by: _____ Date _____
60. Permit Approved for Issue by: _____ Date _____

Date Permit Issued _____	Permit Number _____	Permit and Inspection Fee $ _____

**BUILDING PLAN EXAMINER'S
NOTES AND CORRECTIONS**

Construction Classification

M-Cu. Ft.   (Rate)
Base
Heating
Certif. of Occupancy
Total

**ZONING PLAN EXAMINER'S
NOTES AND CORRECTIONS**

District
Use
Front Yard
Side Yard
Rear Yard
Appeal Sections

Notes

**NOTES ON SUBMISSION TO: HIGHWAY ENGINEER,
TO SANITARY ENGINEER, AND/OR OTHERS.**

[M]

**DEPARTMENT OF
THE BUILDING COMMISSIONER
HAMILTON COUNTY, OHIO
PERMIT APPLICATION FOR
ALL MECHANICAL INSTALLATIONS**

APPLICATION NO.		
ZONE	BOOK	PAGE

Date of Application _____ 19 _____.

APPLICANT — COMPLETE ALL APPLICABLE SPACES ON BOTH SIDES OF THIS FORM. USE INK OR TYPE

1. Address of proposed installation _____
   STREET NAME   STREET NUMBER

   Type of Occupancy _____
                                                          ZIP CODE

   Lot No. _____ Township _____ Municipality or Sec. No. _____

IDENTIFICATION	NAME	STREET ADDRESS	CITY	STATE	ZIP CODE	PHONE NO.
OWNER						
MECH. CONTR.						
GEN. CONTR.						
MECH. PLANS BY						

2. TYPE OF MECHANICAL EQUIPMENT. (Describe & show number of units to be installed)

   A. Furnace ☐ Up Flow ☐ Counter Flow Trade Name _____ Model No. _____ Fuel _____
   B. Boiler ☐ H.W. ☐ Stream ☐ Process Trade Name _____ Model No. _____ Fuel _____
   C. Unit Heater _____ Trade Name _____ Model No. _____ Fuel _____
   D. Comb. Htg. & A.C. _____ Trade Name _____ Model No. _____ Fuel _____
   E. Baseboard Radiation _____ Trade Name _____ Model No. _____ Fuel _____ Input Btuh _____
   F. Incinerator   Trade Name _____ Model No. _____ Fuel _____ Location _____
      Volume of Primary Combustion Chamber _____ Cu. Ft.
   G. Air Cond. Fuel _____ Trade Name _____ Model No. _____ Btuh _____
   H. Heat Pump _____ Trade Name _____ Model No. _____ C. Btuh _____ H. Btuh _____
   I. Unit Ventilator _____ Trade Name _____ Model No. _____ C.F.M. _____
   J. Fan - Type _____ Trade Name _____ Model No. _____ C.F.M. _____
   F. Make-up Air Unit - Type: _____ Trade Name _____ Model No. _____
   L. Storage Tank ☐ Fuel Oil ☐ LP Gas ☐ Waste Oil ☐ Gasoline Cap'y _____ gal. Location _____
   M. Accessory Equipment _____
   N. Automatic Sprinklers _____ Trade Name _____ Area to be Sprinklered _____ Sqft.
   O. Other _____ Trade Name _____ Model No. _____ Fuel _____ Capacity _____
   P. Alterations _____

3. INSTALLED PRICE OF MECHANICAL EQUIPMENT.
   A. Heating $ _____ .00   B. Heating & Air Cond. $ _____ .00   C. Air Cond. $ _____ .00   D. H.V.A.C. $ _____ .00
   E. Storage Tank $ _____ .00, if installed separately or if capacity is 1000 gal. or more.
   F. Other Equipment (describe) _____ $ _____ .00   G. Alterations $ _____ .00

4. DESCRIPTION OF WORK TO BE DONE, LOCATION OF EQUIPMENT, ALTERATIONS, ETC. _____

5. I.B.I. INSPECTION REQUESTED _____ DATE _____

The Undersigned being the Owner, Mech. or Gen. Contractor does hereby covenant and agree to install above noted work in all respects in compliance with the laws of the State of Ohio, and with the Building Code of Hamilton County. Items indicated by checkmark and details on reverse side are included in work covered by Permit.

APPLICANT SIGN HERE _____ ADDRESS _____
(Owner, Mech. or Gen. Contractor) DO NOT WRITE BELOW THIS LINE (OFFICE USE)

☐ Charge included with Bldg. Permit   ☐ Charge Separately

Plan Examiner's Approval: _____ Date _____ 19 _____

If Charged Separately, Date Permit Issued _____ Number _____

SUPPLY DUCT.   Sq.in. Min.
RETURN DUCT.   Sq.in. Min.

Permit and
Inspection Fee $

HEATING & COOLING DATA

IMPORTANT: — COMPLETE ALL APPLICABLE SPACES BELOW.
NEW DWELLINGS: — DATA FOR ALL ROOMS, INC. TOTALS.
REPLACEMENTS: — SHOW TOTALS ONLY.
ADDITIONS: — SHOW EXISTING TOTAL & ADD'N. TOTAL ONLY.

STANDARD USED FOR CALCULATIONS:
☐ ASHRAE
☐ NWAH & ACA
☐ IBR
☐ OTHER (EXPLAIN) _____

5.

Floor	NAME OF ROOM	HEAT LOSS — WARM AIR HOT WATER STEAM (Btuh)	Baseboard Radiation — Electric (Watts)	HEAT GAIN — AIR Conditioning (Btuh)	SUPPLY AIR — HEATING AND/OR Air Cond. (C.F.M.)	RAD-IATION — H.W. OR STEAM (Sqft.)	SUPPLY DUCTS — NUMBER OF OUTLETS PER ROOM	BRANCH AREA (Sqin.)	RETURN DUCTS — GRILLES FREE AREA (Sqin.)	DUCT AREA (Sqin.)
TOTAL								(A)		(A)
					TOTAL BRANCH AREA	(A)	(B)		(B)	
					TOTAL TRUNK DUCT SIZE	(B)				

6. **FURNACE CAPACITY**
☐ NEW ☐ EXISTING
Input _____ Btuh
Output _____ Btuh
At Register _____ Btuh

COMB. HTG. & A.C. CAPACITY
Heating - _____ Btuh Input
Cooling - _____ Btuh

BOILER CAPACITY - H.W. ☐ **STEAM** ☐
Input _____ Btuh _____ Sq. ft.
Output _____ Btuh _____ Sq. ft.
Net Rating _____ Btuh _____ Sq. ft.

Fan Capacity _____ C.F.M. @ _____ INCHES H$_2$O

Size of Combustion Air Opening _____ Sq. in.
(BB-47-30 Ohio Bldg. Code)

Size of Heater Rm. Relief Vent _____ Sq. in.
(BB-7-45 Ohio Bldg. Code)

For Installations other than one, two or three family dwellings, Mechanical Plans must be Submitted in Triplicate with Application.

ABOVE CALCULATIONS MADE BY _____ ADDRESS _____

E

DEPARTMENT OF THE BUILDING COMMISSIONER
HAMILTON COUNTY, OHIO

APPLICATION NO.

SUBMITTAL FOR ENERGY CONSERVATION IN NEW BUILDING CONSTRUCTION

PART I PROJECT INFORMATION

Date of Application _____ 19____.

APPLICANT – COMPLETE ALL APPLICABLE SPACES ON BOTH SIDES OF THIS FORM. USE INK OR TYPE

Address of proposed installation _____ STREET NAME _____ STREET NUMBER _____

Type of Occupancy _____ ZIP CODE _____

Lot No. _____ Township _____ Municipality or Sec. No. _____

IDENTIFICATION	NAME	STREET ADDRESS	CITY	STATE	ZIP CODE	PHONE NO.
OWNER						
SUMITTAL BY						
PLANS BY						
GEN'L CONT'R						

PART II- BUILDING DATA

YES NO
☐ ☐ Residential Building(s)
☐ ☐ Other Building(s)
☐ ☐ Addition

YES NO
☐ ☐ Three (3) Stories, or less
☐ ☐ Floor Area under 5,000 sq.ft.
☐ ☐ Heated
☐ ☐ Cooled

PART III- ANALYSIS METHOD

Select ONE path of compliance   **OR**   Note Exemption

☐ Section 4- Systems Analysis
   See back of this form

☐ Section 5- Component Performance
   See back of this form

☐ Section 6- Acceptable Practice
   Can be used for 1-2-3 family
   residences, & bldgs. that
   qualify under Sect. 601.1.
   USE FORM E6

☐ Bldg. not heated, or cooled

☐ Energy use under 1 watt/sq.ft.
   (3.4 BTU/hr/sq.ft.)

☐ Listed Historic Building

☐ Change of Occupancy in bldg.
   built under this Code. No
   increase in energy demand.
   (submit supporting analysis)

The owner of this building and undersigned, do hereby covenant and agree to comply with all the laws of the State of Ohio and the resolutions of the County of Hamilton pertaining to building and buildings and to construct the proposed building or structure or make the proposed change or alteration in accordance with the plans and specifications submitted herewith, and certify that the information and statements given on this application, drawings and specifications are to the best of their knowledge, true and correct.

Application by _____ Address _____

DO NOT WRITE BELOW THIS LINE (Office Use)

PLAN EXAMINER: _____ DATE: _____
RE-EXAMINED BY: _____ DATE: _____
APPROVED BY: _____ DATE: _____

(SEE OTHER SIDE)

NOTE
Any material, procedure, method, or type of construction, which is acceptable under any other code, standard, or regulation, but not acceptable under the Model Energy Conservation, shall be upgraded to comply with the Energy Code.

PART IV-   REQUIRED REPORTS

    Section 4-Systems Analysis Report, requires an approved computer program analysis, which meets the requirements of Section 402.3, and 402.4 of the Code. This analysis of the proposed design is then compared to a "standard" design. Attach the following,

- [ ] Analysis of annual energy usage, in accordance with Sections 402.3, & 402.4
- [ ] Required computer program analysis, satisfying Sections 402.3, and 402.4.
- [ ] Calculations for comparison with a "standard design" similar building, complying with Section 402.1.
- [ ] All drawings are stamped, or embossed with the seal of a registered (in Ohio) architect, or engineer, who is responsible for the analysis.

ALL INPUTS SHALL BE SHOWN, AND SHALL CORRESPOND TO THE PLANS, AND SPECIFICATIONS THAT ARE SUBMITTED FOR A BUILDING PERMIT

IF SECTION 4 ABOVE HAS BEEN USED, DO NOT FILL OUT SECTION 5 BELOW

    Section 5-Component Performance Approach, is a summary report prepared by the registered design professional(s) responsible for preparing the plans. The report, and the plans, are required to show the following information,

- [ ] Show proposed, and required design information, in tabular form, as shown in the report, and as required in Section 302.1, and 502.1-Tables 5-1 & 5-2.
- [ ] Cross sections on drawings, detailed to substantiate all materials for proposed U_o for floors, walls, ceilings, and roofs.
- [ ] Supporting data for infiltration rates of windows, and doors (502.4)
- [ ] Lists of mechanical equipment, manufacturer, model numbers, ratings, temperature limits, and efficiencies.
- [ ] Air transport factor analysis (503.5)
- [ ] Description and schematic drawings for required HVAC controls (503.8)
- [ ] Insulation specifications for ducts, and piping (503.9, and 503.11)
- [ ] List of equipment for conservation of hot water (504.1)
- [ ] List of electric motors, and equipment
- [ ] Wiring plans showing service voltage, estimated voltage drop, switches, and meters.
- [ ] Lighting report showing lighting load is within lighting power budget (505.3)
- [ ] All drawings are stamped, or embossed with the seal of a registered (in Ohio) architect, or engineer, who is responsible for the report.

ADDITIONAL INFORMATION _____
_____
_____
_____
_____

E6

DEPARTMENT OF THE BUILDING COMMISSIONER
HAMILTON COUNTY, OHIO

APPLICATION NO.

SUBMITTAL FOR ENERGY CONSERVATION— SECTION 6: ACCEPTABLE PRACTICE REPORT

NOTE: This form <u>must</u> be used in conjunction with, and must be attached to a copy of FORM "E". Also, <u>this form can only be used</u> for the following projects; CHECK APPLICABLE CONDITION

☐ RESIDENTIAL BUILDING
(under 5,000 sq.ft. &
3 stories or less)

☐ OTHER BUILDING
(under 5,000 sq.ft.
& 3 stories or less
& HEATED only)

SECTION 602.1— BUILDING ENVELOPE REQUIREMENTS USING CHARTS 6A, and 6B

I. <u>WALLS</u>

    Total exposed wall area (A+)   _____ sq.ft.
    Area of windows, plus ½ area of doors (Aw)   _____ sq.ft.
    Glass % of wall area (Aw/A+ x 100)   _____

    Average U_o for opaque walls   _____
    (Show calculations on separate sheet, if more than
    one type of exposed wall construction is used)
    Estimated U_o from Chart 6A, or 6B   _____
    Maximum permitted U_o from Tables 5-1, or 5-2   _____

II. <u>ROOF/CEILING</u>

    Average U_o of roof   _____
    (Show calculations, on separate sheet, if more than
    one type of roof/ceiling construction is used, or
    if there are skylights)
    Maximum permitted U_o from Tables 5-1, or 5-2   _____

III. <u>FLOORS OVER UNHEATED SPACES</u>

    Average U_o of floor over unheated space   _____
    Maximum permitted U_o of floor from Table 5-1, or 5-2   _____

IV. <u>PERIMETER INSULATION FOR SLAB-ON-GRADE</u>

    R Value of insulation proposed   _____
    R Value of insulation required from Table 5-1, or 5-2   _____

V. <u>INFILTRATION OF WINDOWS, AND DOORS</u>

Manufacturer	Frame Material	<u>Type/Class</u>	*<u>Infiltration Rate</u>
_____	_____	_____	_____
_____	_____	_____	_____
_____	_____	_____	_____
_____	_____	_____	_____

                                  *Submit test results for windows & unweatherstripped doors

V. (cont'd)   CAULKING- By submittal of this form, the Applicant agrees to fully comply with Section 602.3(b), of the Code, in order to prevent air leakage through other building parts, as listed therein.

VI. MECHANICAL EQUIPMENT THERMAL PERFORMANCES
(Heating, Air Conditioning, and Hot Water Service)

Type of Equipment	Mfr. Model No.	Efficiencies Rated Energy Loss Temperature Limits

VII. HVAC CONTROLS

Type	Manufacturer	Model No.	Range	Location(zone)

VIII. DUCT AND PIPE INSULATION

Show, on the plans, the location, type, and thickness of insulation, when req'd. by Sections 503.9, and 503.11 of the Code.

IX. PLUMBING FIXTURES (Shower heads, and public lavatory hot water valves)

Manufacturer	Model No.	Flow Rate (Flow Quantity)

X. ELECTRICAL POWER AND LIGHTING

A. Residential Buildings- Separate meters are required for each dwelling, or apartment unit. Show on plans.

B. Non-Residential Buildings- Complete the Electrical Power and Lighting Budget Report Form (Forms E6-1, and E6-2). CONSULT, AND COMPLY with the requirements of the local electrical inspection agency (copies of E6-1, and E6-2 will be sent to the electrical inspection agency by the Bldg. Dept.)

E6-1

DEPARTMENT OF THE BUILDING COMMISSIONER
HAMILTON COUNTY, OHIO

APPLICATION NO.

ELECTRICAL POWER AND LIGHTING BUDGET FORM- SMALL COMMERCIAL BUILDINGS

NOTE: This form is to be used ONLY for small buildings other than residences which qualify under Section 602.1 of the Code. Attach to "E" & "E6" forms for project

I. POWER FACTOR

   a. Lamps:

      1. ☐ In excess of 15 watts rating-

         A. ☐ Non-inductive (incandescent, etc.)-no special requirements
         B. ☐ Inductive (fluorescent)- limited to 85% min. P.F.

      2. Less than 15 watts rating: no special requirements

   b. Utilization Equipment

      1. ☐ In excess of 1,000 watts rating:

         A. ☐ Non-inductive (resistance heaters, water heaters, etc.)
           ☐ No special requirements
           ☐ With inductive reactance load component;

              1. ☐ Motors (except hoisting, and/or
              2. ☐ Transformers, and/or
              3. ☐ Other: _____
                 (limited to 85% min. P.F. uncorrected or
              4. ☐ Where less than 85%, provided with capacit
                 at the individual equipment item which sha

                   a. be switched with that unit, and
                   b. correct P.F. to at least 90%

              5. ☐ Power Factor Equipment List

FIXTURE OR UTILIZATION EQUIPMENT

Location	Item	Make	Model	Rating	Wattage	Power Factor	Inspector's Use ONLY

II. SERVICE VOLTAGE

                                        1.   2.   3.

    a. Available options: _____ _____ _____

    b. Selected options:

    c. Basis of choice: _____

III. VOLTAGE DROP: (steady state design load condition)

    a. ☐ Branch circuits limited to <u>3.0%</u> max., <u>and</u>

    b. ☐ Feeders limited to <u>3.0%</u> max., <u>and</u>

    c. ☐ Combined total from point of service to any outlet limited to <u>5.0%</u> max.

        Volts x .03=_____AV; x .05=_____AV
        Volts x .03=_____AV; x .05=_____AV

IV. LIGHTING SWITCHING

    ☐ Switching is provided, and shown on the plans, for each lighting circuit, and for portions of each circuit, so that partial lighting required for custodial, or for effective complementary use with natural lighting, may be operated selectively.

E6-2

DEPARTMENT OF THE BUILDING COMMISSIONER
HAMILTON COUNTY, OHIO

APPLICATION NO.

ELECTRICAL POWER AND LIGHTING BUDGET FORM - SMALL COMMERCIAL BUILDINGS

NOTE: This form is to be used ONLY for small buildings other than residences which qualify under Section 601.1 of the Code. Attach this form to "E" & "E6" forms for the project.

SPACE DATA	1	SPACE IDENTIFICATION						
	2	Length L ft (m) / Width W ft (m) — Area A_r ft² (m²)						
	3	Cavity Height ft (m)						
	4	R C R						
TASK 1	5	DESCRIPTION						
	6	Illumination Level, FC						
	7	No. of work Stations / Total Task Area A_1 ft² (m²)						
	8	Luminaire Identification						
	9	Coef. Utilization, CU						
	10	Lamp Efficacy, LE, (lm/W)						
	11	WATTS (TASK 1)						
TASK 2	12	DESCRIPTION						
	13	Illumination Level, FC						
	14	No. of work Stations / Total Task Area A_2 ft² (m²)						
	15	Luminaire Identification						
	16	Coef. Utilization, CU						
	17	Lamp Efficacy, LE, (lm/W)						
	18	WATTS (TASK 2)						
GENERAL	19	Illumination Level, FC						
	20	Area A_g ft² (m²)						
	21	Luminaire Identification						
	22	Coef. Utilization, CU						
	23	Lamp Efficacy, LE, (lm/W)						
	24	WATTS (GENERAL)						
NON-CRITICAL	25	Illumination Level, FC						
	26	Area A_{nc} ft² (m²)						
	27	Luminaire Identification						
	28	Coef. Utilization, CU						
	29	Lamp Efficacy, LE, (lm/W)						
	30	WATTS (NON-CRITICAL)						
	31	WATTS FOR SPACE = [11] + [18] + [24] + [30]						
TOTAL	32	No. of Identical Spaces						
	33	WATTS FOR ALL IDENTICAL SPACES [31] x [32]						

Subtotal this page W.

II. TOTAL LIGHTING WATTS PROVIDED

List quantity, and wattage of fixtures to be utilized:

Quantity   Wattage   Model No.   Manufacturer

_____
_____
_____
_____
_____
_____
_____
_____

HAMILTON COUNTY, OHIO
DEPARTMENT OF THE BUILDING COMMISSIONER

PERMIT APPLICATION FOR: ZONING CERTIFICATE

Initials of Cashier

CASH	CK.	PLANNING APPROVAL	ZONE	BOOK	PAGE

APPLICATION NO.

DO NOT WRITE IN THIS SPACE

APPLICANT — Complete all applicable spaces on this form side **USE BALL POINT PEN OR TYPEWRITER**

(DESCRIBED AS)

1. (a). N S E W side of _____ Street-Zip Code _____ St. No. _____

   (b). _____ feet, N S E W, from intersection of _____ Lot Number _____ Par _____

2. New Subdivision only _____ Township _____ Section No. or Municipality _____

3.

IDENTIFICATION	NAME	STREET ADDRESS	CITY	STATE	ZIP CODE	PHONE NO.
OWNER						
CONTRACTOR						
PLANS BY						

ALL APPLICANTS COMPLETE A THROUGH D

A. TYPE OF IMPROVEMENT
1. ☐ New Building
2. ☐ Addition — Enter number of dwelling units _____
3. ☐ Alteration — Enter number of dwelling units added _____ or deducted _____
4. ☐ Repair, replacement
5. ☐ Parking Lot
6. ☐ Non-Conforming Use
7. ☐ Sign
8. ☐ Other (specify) _____

Describe briefly proposed work _____

D. TYPE OF USE

RESIDENTIAL
11. ☐ One family
12. ☐ Two family
13. ☐ Three family
14. ☐ Four or more family Enter number of units _____
15. ☐ Transient Hotel or Motel ☐ Enter Number of units _____
16. ☐ Accessory Garage ☐ Car Port ☐ Tool Shed
17. ☐ Swimming Pool ☐ Above Gr. ☐ In Gr.
18. ☐ Other _____

NON-RESIDENTIAL
19. ☐ Amusement, recreation, place of assembly
20. ☐ Church, other religious building
21. ☐ Industrial, storage building
22. ☐ Parking Garage
23. ☐ Accessory Garage ☐ Car Port ☐ Tool Shed
24. ☐ Service Station, repair garage
25. ☐ Hospital, institution, nursing home
26. ☐ Office, bank, professional
27. ☐ Public works, utility building
28. ☐ School, college, other educational
29. ☐ Store, other mercantile, restaurant
30. ☐ Swimming Pool
31. ☐ Tank, tower
32. ☐ Other _____

State in detail all existing and proposed uses of this building and premises:

9. ☐ Existing _____

10. ☐ Proposed _____

B. OWNERSHIP
33. ☐ Private
34. ☐ Public (Federal, State, Local)

C. COST (Omit Cents)
35. Estimated cost of improvement for which this application is being made: $_____.00

The owner of this building and undersigned, do hereby covenant and agree to comply with all the laws of the State of Ohio and the Zoning resolution of the County of Hamilton, pertaining to building and buildings, and to construct the proposed building or structure or make the proposed change or alteration in accordance with the plans and specifications submitted herewith, and certify that the information and statements given on this application, drawings and specifications are to the best of their knowledge, true and correct.

Application by _____ Address _____

DO NOT WRITE BELOW THIS LINE (Office Use)

36. Board of Zoning Appeals Case No. _____ 37. Court Case No. _____
38. Zoning Approved by: _____ Date _____
39. Zoning Certificate: _____ Date _____
40. Approved for Issue by: _____ Date _____

Date Zoning Certificate Issued _____ Certificate Number _____ Certificate and Inspection Fee $ _____

**BUILDING PLAN EXAMINER'S
NOTES AND CORRECTIONS**

Construction Classification

M-Cu. Ft. (Rate)
Base
Heating
Certif. of Occupancy
Total

**ZONING PLAN EXAMINER'S
NOTES AND CORRECTIONS**

District
Use
Front Yard
Side Yard
Rear Yard
Appeal Sections

Notes

**NOTES ON SUBMISSION TO: HIGHWAY ENGINEER,
TO SANITARY ENGINEER, AND/OR OTHERS.**

Department of the Building Commissioner
Hamilton County, Ohio
Room 312, Temple Bar Building
138 E. Court Street
Cincinnati, Ohio 45202

**DEPARTMENT OF
THE BUILDING COMMISSIONER
HAMILTON COUNTY, OHIO**

APPLICATION FOR FINAL INSPECTION AND CERTIFICATE OF OCCUPANCY

APPLICANT:
Complete items 1 thru 9.
Use ballpoint pen or typewriter.

F

1. Application Number _____ Building Permit Number _____

2. (a) N S E W side of _____ Street-Zip Code _____ St. No. _____
   (b) _____ feet, N S E W, from intersection of _____ Lot Number _____ Par _____

3. New Subdivision only _____ Township _____ Section No. or Municipality _____

4.
IDENTIFICATION	NAME	STREET ADDRESS	CITY	STATE	ZIP CODE	PHONE NO.
OWNER						
BUILDER						
DRAWINGS BY						

5. Application For: ☐ New Building ☐ Addition ☐ Alteration

6. Occupancy: (Check One)
   ☐ 1 Family ☐ 2 or 3 Family ☐ Multiple Res. Building ☐ Hospital or Home
   ☐ School ☐ Place of Assembly ☐ Business ☐ Industrial ☐ Special Occupancy
   _____ Other

7. I/We certify that I/We am/are the Builder/Builders of the above described improvement and I/We will request FINAL INSPECTION of said improvement.

8. I/We hereby certify that the above described improvement will substantially comply with all provisions of the Hamilton County Building Code and/or Ohio Building Code and its amendments and that there will not be any areaway apron drains, foundation tile drains, downspouts, or storm water drains of any type connected to the Sanitary Sewer System.

9. I/We the Builder/Builders understand that any violation of the Hamilton County Building Regulations; any false information on this application; and any occupancy before final inspection has been made and a Certificate of Occupancy is issued make me/us subject to penalties provided in the Hamilton County Building Code, Section A-14.

_____   _____
Builder/Builders   Signature (Owner of Firm or Officer of Corporation)

_____   _____
Date   Title

* DO NOT WRITE BELOW THIS LINE *   (OFFICE USE)

Date of Final Building Inspection _____ BY _____

Date of Final Plumbing Inspection _____ Plumbing Permit No. _____

Date of Final Electric Inspection _____ Plumbing Contractor _____

Date of Certificate of Occupancy _____

Comments: _____

**DEPARTMENT OF THE BUILDING COMMISSIONER
HAMILTON COUNTY, OHIO**
312 TEMPLE BAR BLDG., 138 EAST COURT STREET
CINCINNATI, OHIO 45202
PHONE: 632-8362

PLAN EXAMINERS' CHECK LIST
BOXES CHECKED SHOW ITEMS OF INFORMATION NOT COVERED ON SUBMISSION DRAWINGS, WHEN IN FACT THEY ARE REQUIRED BY BUILDING OR ZONING CODE*.

APPLICATION NO.	
APPLICANT:	
ADDRESS:	
CITY: _____ STATE: _____ ZIP: _____	
PROJECT:	
TOWNSHIP:	
STREET & NO.:	

APPLICATIONS

#	
1	FILL OUT FORM "____" FOR _____
2	INDICATE ON EACH SHEET OF DRAWINGS, NAME OF DESIGNER, OWNER, CONTRACTOR & LOCATION OF JOB.
3	MAKE CORRECTIONS TO TRACINGS & RE-SUBMIT NEW PRINTS: NOT ACCEPTABLE FOR PLAN REVIEW.
4	NOTE ITEMS CHECKED ON SURVEY CHECKLIST ON REVERSE SIDE.
5	COMPLETE REVERSE SIDE OF MECHANICAL APPLICATION FORM "M".
6	
7	
8	

STRUCTURAL

#	
25	SHOW PROPER FOUNDATION WALL DESIGN FOR BACK FILL CONDITIONS.
26	SHOW PROPER FOOTING SIZE AND SLAB THICKNESS.
27	SHOW DESIGN, SECTION & DETAILS OF RETAINING WALLS.
28	SHOW MATERIALS, SIZES, BEARING, ANCHORAGE OF BEAMS, HEADERS, COLUMNS, LINTELS, ETC.
29	INDICATE ROT OR TERMITE PROTECTION TO EXTERIOR OR BELOW GRADE.
30	SHOW MATERIALS, SIZES, DIRECTION OF FLOOR, ROOF FRAMING & TYPE OF FLR. BRIDGING.
31	SHOW TRUSS DESIGN DATA OR MODEL, FABRICATOR, TYPE & MANUFACTURER.
32	

CONSTRUCTION STANDARDS

#	
9	SHOW BUILDING CROSS SECTIONS; ALL HEIGHT DIMENSIONS, ALL EXTERIOR ELEV'S & WALL SECTIONS.
10	SHOW FIREPLACE CONST., VERTICAL SECTION, DETAILS & PLAN.
11	SHOW STAIR SECTION WITH DIMENSIONS, AND HEAD ROOM CLEARANCE.
12	SHOW RAILINGS ON STAIRS (INSIDE, OUTSIDE)
13	SHOW GLASS TYPE & THICKNESS FOR LIGHTS OVER 7.3 SQ. FT. IN AREA.
14	SHOW ACCESS TO LIVING UNIT, CRAWL SPACE, OR ATTIC; SHOW SECONDARY EGRESS.
15	ROOM OR ROOMS LESS THAN MINIMUM SIZE, OR HAVE LESS THAN MINIMUM PRIVACY.
16	SHOW WINDOW TYPES, SIZES, AND OPERABLE PARTS.
17	SHOW ATTIC AND, OR CRAWL SPACE VENTILATION.
18	SHOW VENTILATION OF TOILET ROOM.
19	SHOW INSULATION (R-FACTORS) OR FILL IN INSULATION FORMS (2 REQ'D.)
20	

MECHANICAL

#	
33	SHOW LOCATION OF HEATING DEVICES & FLUES, OR CHIMNEYS.
34	SHOW PROPER HEIGHT OF CHIMNEY ABOVE ROOF.
35	SHOW PROPER COMBUSTION AIR INTAKE, IF REQUIRED.
36	SHOW DUCT INSULATION WHERE REQUIRED.
37	DETAILS AND SECTIONS REQUIRED FOR DUCT WORK UNDER CONC. SLAB.
38	SHOW LOCATIONS OF DISCHARGES OF TOILET FANS AND KITCHEN EXHAUST FANS.
39	SHOW DETAILS AND SECTION OF CISTERN.
40	SHOW LOCATION OF FUEL TANK (OIL, L.P.)
41	SHOW LOCATION OF AIR CONDITIONING COMPONENTS.
42	
43	
44	

FIRE RESISTANCE

#	
21	SHOW FIRE SEPARATION BETWEEN GARAGE & REST OF DWELLING, BEAM, PLASTIC PIPE & COL.
22	SHOW PROTECTION OF COMBUSTIBLE MATERIALS ADJACENT TO HEATING DEVICE OR CHIMNEY.
23	
24	

OTHER

#	
45	
46	
47	
48	

DEPARTMENT OF THE BUILDING COMMISSIONER
HAMILTON COUNTY, OHIO
312 TEMPLE BAR BLDG., 138 EAST COURT STREET
CINCINNATI, OHIO 45202
PHONE: 632-8362

SITE PLAN CHECKLIST
BOXES CHECKED SHOW ITEMS ON SURVEY-SITE PLAN NOT COVERED ON SUBMISSION DRAWINGS, WHEN IN FACT THEY ARE REQUIRED BY BUILDING OR ZONING CODE*.

INFORMATION REQUIRED TO BE ON THE SURVEY AND/OR SITE PLAN

1. ☐ Title, Name of Builder and/or Owner, Date, Name of Surveyor and Seal.

2. ☐ Indication of County, Township, Section Number, Parcel Number, Lot Number.

3. ☐ North Arrow, Scale Indication (not smaller than 1" = 50')

4. ☐ Indication of property lines, dimensions, bearings, monuments found and/or set. Give lot area.

5. ☐ Adjacent lots within 100 ft. (indicate if vacant) Adjacent buildings within 50 ft.

6. ☐ Indication of streets, roads, alleys including widths, type of paving, centerlines, and dedication or not.

7. ☐ Distance of lot from nearest intersecting street.

8. ☐ Easements for utilities, access drives or other.

9. ☐ Driveways, parking aprons, access sidewalks around buildings, indicate all materials.

10. ☐ Proposed buildings, accessory structures, walls, fences with complete dimensions.

11. ☐ Yard setbacks: Front, sides, rear yard.

12. ☐ Indicate existing and proposed grades at:
    a. Property corners.
    b. Corners of proposed structure.
    c. Minimum of two (2) points on the side lot lines opposite the corners of the proposed structure.

13. ☐ Elevations of all floor levels of building.

14. ☐ Sewers, manholes, house lateral, with type of sewer, size of sewer, invert elevations at manholes and house lateral connection.

15. ☐ Indication of storm water disposal: downspouts, drain tile, yard drains, area well drains. Show swales and drainage courses.

APPLICATION NO.	PERMIT NO.

CLARIFICATION MEMO – RESIDENTIAL

The following items are resolved, with the understanding that the items must be provided as noted hereon, and in accordance with the Hamilton County Building Code. (Note checklist items with asterisk*)

_____
_____
_____
_____
_____
_____
_____
_____
_____
_____
_____
_____
_____
_____
_____

Date _____ Applicant _____

Examiner _____

DEPARTMENT OF THE BUILDING COMMISSIONER
HAMILTON COUNTY, OHIO
312 TEMPLE BAR BLDG.   •   PHONE 632-8362
138 EAST COURT STREET   •   CINCINNATI, OHIO 45202

THIS PROJECT REQUIRES APPROVAL OF THE SYCAMORE TOWNSHIP FIRE SAFETY BUREAU BEFORE A BUILDING PERMIT CAN BE ISSUED. CALL 791-8423

THE APPROVAL OF PLANS AND APPLICATION DOES NOT PERMIT THE VIOLATION OF ANY SECTION OF THE BUILDING CODE. Separate Applications shall be submitted for permits for buildings and structures, plumbing heating, air conditioning, sprinklers, signs, fences, oil tanks, and all other facilities or construction as required by the Code.

DEC 19 1980

Ralph W. Liebing
Building Commissioner

DEPARTMENT OF THE BUILDING COMMISSIONER
HAMILTON COUNTY, OHIO
312 TEMPLE BAR BLDG.
138 EAST COURT STREET
CINCINNATI, OHIO 45202
PHONE 632-8362

BUILDING PERMIT

THIS PERMIT MUST BE DISPLAYED DURING CONSTRUCTION

OWNER	APPLICATION NO.
ADDRESS	
PERMIT FOR	
OCCUPANCY	
ADDRESS	

LOCATED ON ___ SIDE OF STREET ___ FEET FROM ___ SITUATED IN ___

TOWNSHIP ZONED ___ SECTION ___
LOT ___ BOOK ___ PARCEL ___

ARCHITECT OR ENGINEER ___
CONTRACTOR ___

1. ELECTRICAL WORK REQUIRES PERMIT AND INSPECTIONS MADE BY INSPECTION BUREAU, INC. 2. PLUMBER NEEDS GREEN COPY OF THIS PERMIT FOR PLUMBING PERMIT. 3. MECHANICAL CONTRACTORS MUST OBTAIN SEPARATE PERMIT FROM THIS DEPARTMENT IF WORK IS NOT INCLUDED IN THIS PERMIT.

THE ISSUANCE OF THIS PERMIT DOES NOT ALLOW THE VIOLATION OF THE BUILDING CODE OR OTHER GOVERNING REGULATIONS.

PERMIT FEE	CUBIC FT. CONT.	EST. COST OF WORK	APPROVED BY	ISSUED BY
ZON. $		$		
BLDG. $				
MECH. $	PERMIT NUMBER	VOID UNLESS STAMPED "PAID" AND WORK IS STARTED WITHIN SIX MONTHS		
C.O. $	C 10821			
TOT. $				

N O T I C E
MANDATORY INSPECTIONS

HAMILTON COUNTY BUILDING CODE, SECTION A-9(B):

B. NOTICE. It shall be the responsibility of the holder of a permit to notify the Building Official when work is ready for the various inspections required by the terms of the permit or the approved rules. Such notice shall be given within a reasonable time within which such inspection is desired but in no event shall be less than 24 hours. Notice given on a Friday or on a day prior to a legal holiday shall not constitute notice for inspection on a Saturday, Sunday or holiday unless arrangements have been made under approved rules for inspection on such days. Before giving such notice the holder of the permit shall first test the work and satisfy himself that it conforms to the approved plans and specifications and the requirements of this Code.

AGENCY		INSPECTION
(A)	1.	*Soil Inspection (before pouring concrete)
(B)	2.	*Rough Plumbing
(A)	3.	*Framing Inspection (framing complete, but not covered)
(A)	4.	Exterior Cladding
(A)	5.	Gutters and Downspouts
(C)	6.	*Rough Electric
(A)	7.	*Insulation (after rough elec. before interior finish is applied)
(A)	8.	*Heating and Air Conditioning
(A)	9.	Interior Finishes
(C)	10.	*Final Electric
(B)	11.	*Final Plumbing
(A)	12.	Backfill and Grading
(D)	13.	*Driveway Grading Inspection
(D)	14.	*Driveway Completion
(A)	15.	*Final Inspection for Certificate of Occupancy (NO OCCUPANCY PRIOR TO THIS INSPECTION)
(A)	16.	*Swimming Pools (after excavation is completed)
(A)	17.	*Fences and Signs (when completed)
(A)	18.	*Final inspection for Pools (Fences, constr. & elec.verification)

* Denotes "called-for" inspections (mandatory)

KEY TO INSPECTION AGENCIES:

(A) Dept of the Bldg Commissioner
    Room 312 Temple Bar Building
    Telephone: 632-8362
               632-8643 (after 4 p.m.)
    Inspectors in Office 8-9 a.m. daily

(B) Plumbing Department
    Hamilton County Board of Health
    Room 1010 Temple Bar Bldg
    Telephone: 632-8455

(C) Inspection Bureau, Inc.
    Room 503C Alms & Doepke Bldg
    222 E. Central Parkway
    Telephone: 381-6080

(D) Hamilton County Engineer
    Room 800 Temple Bar Bldg
    Telephone: 632-8500

NOTICE:

1. When floor plans and/or building structure has been changed from the approved drawings, new drawings and permit application (marked "revision") must be submitted to the Building Commissioner's Office for approval.

2. When mechanical equipment has been changed or modified, a new mechanical form must be submitted to the Building Commissioner's Office. New form shall be marked "revision".

DEPARTMENT OF THE BUILDING COMMISSIONER
    HAMILTON COUNTY, OHIO

                    THE CONCEPT, AND ATTITUDE OF
                    FIELD INSPECTION

The following are simple ideas regarding field inspection, and the attitude necessary to be an effective field inspector.

1. Yours is a job of checking for compliance with the code, <u>and more importantly,</u> with the drawings submitted for approval. The approval of the drawings, and other documents, is the basis of the permit.

2. <u>You are expected to see</u> that the structure is built in accord with the APPROVED documents. By utilizing the approved project drawings specifications, and departmental checklists, <u>during each and every inspection,</u> you will be able to cover, and inspect all critical areas. WHEN YOU GO ONTO THE SITE, THE DOCUMENTS GO WITH YOU!

3. If something on the job is different from the drawings, ALWAYS question it, and check it fully before resolving it in your mind.

4. Some items may vary during construction, and you may draw on your experience as to whether or not the change is proper, safe, etc. Any major change must be stopped until you can consult with the plan examiner, since the full impact of the change may not be apparent in the field.
    DO NOT assume it is the correct answer, and fully proper

    DO NOT take responsibility unduly by approving it without consultation

    REMEMBER, the project people (owner, designer, builder) are to show us how they intend to comply with all of the regulations. We are not a design agency, and have no responsibility for how the project complies; we are there to see that it does comply, in one manner or another.

5. Be open to new materials, construction methods, techniques, etc., but don't accept them blindly. CHECK THEM OUT!

6. Remember, you have the full expertise of the department staff to back you up; USE IT!. Use your experience and check with others, i.e. supervisors, plan examiners, zoning examiners, etc. Use the technical information in the office, or ask for such information from the project people, if, it is not in the office.

7. NEVER impose your opinion, your will, your method, or your "way" on the the project. Be sure you have the code, or technical data to back you up.

8. AT ALL TIMES, be courteous, positive, helpful (but firm), and approach your work in a sincere, professional, workman-like manner.

LIGHT WOOD FRAMING STANDARD FIELD CHECKLIST

NOTE: This is a reminder list or guide ONLY, and is not to be used for any other purpose. It covers the most checked light framing items, but does not pretend to be complete. Contact your supervisor, or plan examiner for specific problems, and assistance.

MATERIAL-

Check grades of lumber, etc. against the plans, and specs.

Check field delivered material for markings of grade (such marks are required)

Check for splits, rot, fungus, termites, warpage, knots, and other damage

Is material required to be preservative treated, or impregnated for any reason?

Check plywood as well as framing material

Check intended hangers, bolts, plates for proper size, gauge, and coatings

Check prefab materials, such as stress-skin panels, fabricated girders, joists, gang-nail trusses, for grade, and manufacturer. Do they match plan information?

What quality control at point of manufacturer?

ERECTION AND FRAMING-

Check basic member sizes, and spacing, Check span, notching, hole boring, bridging (if required), and bearing

Are hangers, tiedowns, or special attachments required? If so, are theyproperly done

Is nailing per code schedule, or per design?

Has nailing caused end splitting, etc.? Is nailing well-seated, and well executed?

Are base plates anchored per code? Sill plates?

Has termite treatment of underfloor area been completed? Certificate?

Has framing been damaged due to work of other trades? Check this continuously!!!!!!!

Are members doubled as required?

Is crawl space properly vented?

FIRESTOPPING- check for proper installation in all required locations

Check clearances to vents, and heat producing appliances

Are porch members treated?

Is sheathing nailed as per schedule, code, or manufacturer's instructions

Is additional lateral, or let-in bracing required?

Are roof truss tie-downs as required by code?

Do bottom plates have fully, 100% bearing

Check room sizes, ceiling heights, headroom clearances, floor area

Check size of windows; check plan, and actual installation with code

Check header sizes

Check end bearing of lintels

Check strike bracing for burglar security

Check stairs for net depth of stringer, and firm attachment; check riser and tread sizes; headroom, rails

Are attic areas required to be divided?

Are dormers, or special openings framed correctly; necessary "doubling" of members?

Is attic access provided, size to meet code? proper fire rating?

Are attic vent areas provided, and properly sized?

Is there proper clearance between roof framing and combustibles?

Do intersecting walls attach properly

Are large floor cutouts edge blocked

Roof sheathing; proper bearing, nailing, grade, thickness; know APA grade markings

Are plywood roof deck clips required?

Especially check window walls, and large openings for structural bearing above

Are joints in built-up members staggered; properly nailing, spiked, or bolted

If adhesive floor attachment is used, check application, temperature, and conditions at time of application

Wood siding, check required thickness, nailing, backing, waterproof quality, joining, and clearance from finish grade

Recheck vents, fireplaces, and chimneys (including prefab units) for clearance to combustibles, firestopping, etc.

Check exposed surfaces of raised, galvanized column supports

Check exterior stair, porches, patios, etc., for required railings, yard dimensions

CHECKLIST FOR INSULATION INSPECTION

NOTE: This is a reminder list, or guide ONLY, and is not to be used for any other purpose. It covers the most checked items, but does not pretend to be complete. Consult your packet of details for further information. Contact your supervisor, or plan examiner for specific problems, and assistance.

GENERAL-

Be sure that you inspect ALL insulated areas; walls, floors, ceilings, etc.

Verify all information on the drawings, to insure that the proper material is being used in the proper places.

If you cannot account for the material being used (R listing on the wrapper, etc., make the builder supply the information)

Understand that because the insulation is not as thick as the framing member does not necessarily mean there is something faulty: simply make sure the proper thickness of insulation is present.

Watch for "over-insulation", where the material must be compressed to allow the other materials to be installed. This compression is a negative factor and reduces the insulating value of the material.

FLOORS-

Buildings with slab-on-grade construction require perimeter insulation- be sure that there is a positive barrier (insulation) between the outside and inside. Check details

At the sill, make sure that the header is insulated if the insulation in the floor joists does not cover the entire header.

Check for a careful fit around bridging

Heating cut-outs in floors should be neatly cut(not undersized), and are usually cut out the full size of the partition stud space.

WALLS-

Make sure all of the small cracks, and spaces in the wall system are fully insulated. (between wall panels, around door and window frames, etc.

The walls around a stair leading from an unheated basement to the first floor MUST be insulated. ALSO, the underside of the stairs to the second floor, or the ceiling of the stairhall MUST be insulated.

A duct in a wall between a heated and unheated space, must be insulated. If the duct is sized to fill the entire stud space, the insulation must be applied to the face of the wall, on the "cold" side.

At the second floor "sill", insure that the rim joist (or edge joist, or band board) is properly insulated.

At windows and doors, insure that the vapor barrier covers the insulated space between the finished frame and the rough framing.

Holes in sheathing boards must be patched, and fully closed.

Check for a continuous vapor barrier; watch joints for proper overlapping of plastic, or flanges of the insulation wrappers.

Insure that insulation is installed: behind electrical boxes
behind pipes or conduit in walls
behind bath tubs, and shower units.

Check windows and doors; are they the ones listed on the drawings?

CEILINGS-

Ceiling insulation must be visually checked; it may be tough, BUT CHECK IT.

Insulation should extend over the top plates of the walls.

Check for baffle, or retainer at walls, so loose (blown-in) insulation has not covered the eave vents.

Combination of batt, and blown-in insulation is acceptable; however, this system should be shown on the drawings- if not, YOU verify it.

Insure that insulation is at least 3" from all recessed light fixtures; an insulating hood, loosely fit over the fixture must be provided. Check with plan examiners for details.

Insure that voids between framing and chimneys is insulated, with non-combustible insulation.

MISCELLANEOUS-

Building overhangs, no matter how small, must be insulated.

Really check insulation of areas with a lot of piping, or wiring run in them. The insulation must be fully, and properly installed -- NO VOIDS

Ducts run in unheated areas must be insulated- check details

Watch areas with various floor levels, partial basements, etc. Make sure there is a continuous pattern of insulation in the detail.

Bulkheads (dropped ceilings) over kitchen cabinets should not be left open to the attic above. Run the drywall through to cut off the area, and allow for insulation.

Caulking must be checked for full continuous beads around all window frames, door frames, and other openings.

Check for air infiltration data on doors and windows.

Large voids between wood sill plates and tops of foundations should be filled to prevent air infiltration.

Watch exterior joints for tightness, sealing, etc. Again for infiltration.

Jobs with foam-in-place insulation which you can't see, require thermographs. No foam insulation of any type can be left exposed, anywhere in the building.

DEPARTMENT OF THE BUILDING COMMISSIONER
HAMILTON COUNTY, OHIO

<u>BUILDING INSPECTION RECORD</u>
<u>OHIO BASIC BUILDING CODE OCCUPANCIES</u>

Inspector shall indicate, next to each inspection item, the date of his inspection, his initials, information required, or N.A. if item is not applicable. When complete, the original shall be attached (stapled) to the Inspector's copy of the permit, and kept with the permanent office records.

1. Type Occupancy _____
   Application No. _____ Permit No. _____ Date Issued _____
   Project Address _____

2.   Date   Insp   ☐ Bearing Soil and Location Inspection
   _____   Building location per survey
   _____   Revised survey requested
   _____   "   "   submitted
   _____   "   "   approved
   _____   Bearing soil approved
   _____   Copy of soil test report on file
                           (If fill, have soil test reports been received?)
   _____   Floor elevations

3.   ☐ Foundation
   _____   Recheck building location
   _____   Foundation per drawings
   Comments: _____
   _____

4.   ☐ Framing (Use supplemental checklist before filling in below)
   _____   Floor framing per drawings/standard
   _____   Wall framing per drawings/standard
   _____   Roof framing per drawings/standard
   Comments: _____
   _____

5.   ☐ Masonry
   _____   Masonry per drawings/standard
   _____   Firewalls per drawings/code/standard
   _____   Masonry bond
   _____   Mechanical bond
   _____   Bearings
   Comments: _____
   _____

Page Two -   O.B.B.C.

6.   Date   Insp   ☐ Insulation (Use supplemental checklist before filling
                               in below)
     _____   Floor insulation per drawings/energy submittal
     _____   Wall insulation per drawings/energy submittal
     _____   Air infiltration
     _____   Ceiling/roof insulation per drawings/energy submittal

7.   ☐ Backfill & Grading
     _____   Backfill per drawings
     _____   Damp-proofing foundations
     _____   Footing drain tile

8.   ☐ Roof system
     _____   Roof deck per drawings
     _____   Insulation per drawings
     _____   Covering per drawings or specs
     _____   Flashing
     _____   Gutters & Downspouts
     _____   Downspout discharge
     _____   Scuttles, other roof access

9.   ☐ Ceiling heights
     _____   Rooms
     _____   Corridors
     _____   Stairways

10.   ☐ Exit facilities
     _____   Doors, including appropriate label, if required
     _____   Hardware
     _____   Exit Signs
     _____   Stairways
     _____   Illumination

11.   ☐ Fire-Compartments & Fire-Stopping
     _____   Partitions & joist spaces
     _____   Openings around pipes, conduits, ducts
     _____   Furred wall (combustible furring)
     _____   Concealed spaces
     _____   Attic areas
     _____   Fire walls
                                Other

Page Three -   O.B.B.C.

12.   Date   Insp   ☐ Fire Resistant Rated
    _____   Partitions
                                   Ceilings including hold-down clips and light
    _____   fixture protection
                                   Doors and frames, U.L. label & appropriate
    _____   inspection decal

13.   ☐ Interior finish and trim
    _____   Required certificates

14.   ☐ Decorative materials
    _____   Required certificates

15.   ☐ Carpet
    _____   Required certificates

16.   ☐ Openings
    _____   Doors per plans: caulking
    _____   Windows per plans: caulking
    _____   **Energy requirements met**
    _____   Safety glass per plans

17.   ☐ Railings
    _____   Handrails at stairs
    _____   Platform guard rails

18.   ☐ Toilet Rooms
    _____   Male
    _____   Female
    _____   Number of fixtures per plan

19.   ☐ Emergency lighting system
    _____   Operative

20.   ☐ Fire Extinguishers
    _____   Type and locations, per plan
    _____   Mounting height

21.   ☐ Sprinkler system
    _____   Matching thread connections at F.H.
    _____   Flushing inspection (interior)
    _____   Drain inspection
    _____   Alarm inspection
    _____   Witness final test; copy of certificate in file
                                   Flushing inspection (exterior)

Page Four - O.B.B.C.

22.   Date   Insp   ☐ Smoke detectors
         _____   Type and location
         _____   Operative

23.   ☐ Fire alarm system
         _____   Operational

24.   ☐ Mechanical Work
         _____   Type of System
         _____   Fuel Supply
         _____   Equipment per drawings
         _____   Equipment per energy submittal
         _____   Duct sizing & construction
         _____   Fire dampers as required
         _____   Combination air & relief vent
         _____   N.F.P.A. requirements
         _____   Type of fire protection
         _____   State boiler inspection

25.   ☐ Kitchen Equipment
         _____   Hood, duct, fan: per Code
         _____   Fire Suppression System

26.   ☐ Tanks
         _____   Type: Fuel oil
         _____   L.P.
         _____   Gasoline
         _____   Other
         _____   Size and location per plans
         _____   Protection per plans

27.   ☐ Site
         _____   Paving
                                         Type and location per plans
         _____   Driveways
                                         Location per plans
         _____   Drives
                                         State and/or County approval
         _____   Wheel stops per plans
                                         Islands
         _____   Location & type per plan
                                         Signs on building
         _____   Size and location per drawings
                                         Ground signs
         _____   Location and type per plans

Page Five -   O.B.B.C.

27.   Date   Insp   ☐ Site (Continued)
                                            Sidewalks
_____   Location per plans
                                            Fences
_____   Location, type & height per plans
                                            Guard rails
_____   Type & location per plans
                                            Catch basins, trench drains, drainage
_____   Location per plans
                                            Lighting
_____   Illumination & direction
                                            Parking
_____   Stripping layout & car count per plan
                                            Landscaping, Planting & Greenbelts
_____   Type & location per plans
                                            Guard posts
_____   Location per plans
                                            Retaining walls
_____   Location per plans
                                            Fire protection lines & F.H.
_____   Size & location per plans

28.   ☐ Elevator and Escalator
_____   State inspection made
                                            Certificate posted in cab

29.   ☐ Board of Health
_____   Sanitarian approval

30. ☐ Electric:
      Rough: _____(Date)   Final: _____(Date)

31. ☐ Plumbing:
      Rough: _____(Date)   Final: _____(Date)

32. ☐ Township Fire Inspection:
      Rough: _____(Date)   Final: _____(Date)

33. ☐ Inspector & Permit History:
      Original Inspector   _____   Date Assigned   _____
      Subsequent Inspectors   _____   Date Assigned   _____
                              _____   Date Assigned   _____
      Permit extensions   _____   Issued   _____
                              _____   Issued   _____
      Permit invalidated   _____   (Date)

Page Six -   O.B.B.C.

34. ☐ Permits required other than Building Permit:

    Heating   # _____   Issue date: _____

    Air Conditioning   # _____   Issue date: _____

    Electric   # _____   Issue date: _____

    Sprinkler   # _____   Issue date: _____

    Fence   # _____   Issue date: _____

    Tank   # _____   Issue date: _____

    Sign   # _____   Issue date: _____

    Other   # _____   Issue date: _____

                          # _____   Issue date: _____

35. ☐ No-Work Progress:

    Date of Inspection   _____   Inspector   _____

    "   "   "   _____   "   _____

    "   "   "   _____   "   _____

    "   "   "   _____   "   _____

    "   "   "   _____   "   _____

    "   "   "   _____   "   _____

36. ☐ Written orders issued

    Type: _____   Date: _____   Insp: _____

    "   _____   "   _____   "   _____

    "   _____   "   _____   "   _____

    "   _____   "   _____   "   _____

    "   _____   "   _____   "   _____

    "   _____   "   _____   "   _____

    "   _____   "   _____   "   _____

    "   _____   "   _____   "   _____

37. ☐ Approved spaces:

    Quantity of dwelling units   _____

    Quantity of Tenant spaces   _____

    Quantity per drawings   _____

38. ☐ Approved for conditional occupancy:

    Date _____   Inspector _____

39. ☐ Approved for final Certificate of Occupancy:

    Date _____   Inspector _____

Page Seven -   O.B.B.C.

40. ☐ Approved by Supervising Inspector:

   Date _____   Inspector _____

41. ☐ Miscellaneous Comments:

_____
_____
_____
_____

DEPARTMENT OF THE BUILDING COMMISSIONER
HAMILTON COUNTY, OHIO

Certificate Number _____

Permit Application Number _____

Township _____

CERTIFICATE OF OCCUPANCY AND USE

ISSUED TO _____

_____

* * * * * *

THIS IS TO CERTIFY that the improvement erected at: _____

_____

for use as _____
HAS BEEN INSPECTED by the Department of the Building Commissioner, and that the applicable Building Code Regulations have been substantially complied with and the same may be used or occupied as set forth herein.

_____
Building Inspector

Date _____   _____
                                                                                    Building Commissioner

IMPORTANT PAPER — KEEP WITH DEED & PROPERTY PAPERS

INDEX

Addenda, 35
Administrative superintendent, contractor's, 42
Adversary position/condition, xii

Building codes:
   effect on costs, 66
   inclusion by experience, 55, 63
   history of, 53
   minimum standards set by, 54, 61
   necessity of inspection, 53–55
   as part of design sequence, 55
   purpose of, 65
   reference standards in, 58
Building construction:
   abuse of, 74
   attention to detail of, 54
   negated by use/occupants, 54, 72, 74
Building department:
   attitude of, 68
   checklists, 101–115
   forms, 77–116
   organization of, 59
   records, 67
Building inspection:
   as code officials system, 55
   checklists, 101–115
   concept of, 100
   governmental, 53–55, 58
   list of, 99
   material test data, 60
   need for, xi, 40, 53, 55
   on-going sequence, xii, 33, 62
   terminal documents, 75, 116
   timely, 10, 16, 34
   see also, Construction inspection; Inspection system
Building inspectors:
   attitude of, 63–64
   basic interest of, 55
   certification of, 62
   characteristics of, 61–62
   continuing education for, 64
   education of, 62
   professionalism, 64
   techniques of operation, 62–65

Building materials:
   equivalency of, 60
   test data for, 60
Building permit:
   application for, 80–94
   issuance after plan review, 55
   sample form of, 98

Change orders, 35
Clerk of works:
   as added service of design professional, 20
   as ancillary of construction management, 29
Communications:
   avoid misunderstandings, 5
   establishing goal of design concept, 5
   establishing roles and modes of operation, 5
   failure in, 7, 31, 33
   as feedback, 5
   guidelines for, 7
   proper channels for, 7
Construction drawings:
   as basis for inspection when approved, 59
   as function of building codes, 53, 55
Construction inspection:
   as evidence of performance, xii, 36, 42
   on-going sequence of, xii, 33, 62
   see also, Building inspection; Inspection system
Construction management, as system, 22
   history of development, 22
   prior experience in, 23
   single purpose department for, 23, 27
Construction manager:
   background of, 2, 22, 23
   contract requirements, 27
   design professional as, 20
   duties specified- AIA, 23
   implied inspections, 23
   owner's selection of, 28
Construction meetings:
   periodic/progress, 14, 15
   pre-construction, 14
   pre-installation, 48
Consultants:
   design professional's, 21
   representation on job site, 21

Index

Contractor:
    attitude of, 3, 40, 45, 52
    capabilities of, 3, 46, 47
    superintendent for, 42
Contracts:
    construction manager's, 27
    contractor's, 35
    design professional's, 27
    establish representation, 33
    parties to, 31, 44, 46–50
    subcontractor's, 44, 46
    traditional configuration of, 44, 56

Design professional:
    attitude of, 9, 63
    as owner's agent, 9, 10, 30
    as project coordinator, 11
    responsibilities in general, 9, 10, 16, 58
    responsibility similar to building inspector, 65
    services beyond contract, 12, 20
    supervision/inspection by, 9, 13, 20

Education:
    building inspector's, 62
    contractor's, 47
    design professional's, 20, 35
    of future citizens, 68, 76
    lack of public, 68, 69
    subcontractor's, 47, 49
Examination, as construction inspection, xi

Fast tracking, as system, 1–3
Federal control, 75
Feedback:
    as communications, 5
    list of features, 6
    need for proper, 5
Field/job conditions, 4, 34, 60
Fire protective measures:
    between buildings, 54
    within buildings, 54

General Conditions:
    as applies to construction manager, 23
    of Contract for Construction, 23
    as design professional's tool, 24
    example of Supplementary, 39

Inspection agencies:
    governmental, 53–55, 58
    in inspection system, xii
    limitations of, xii
    joint inspections by, 41, 53
    team effort of, xii

Inspection departmentalization:
    for construction management, 20, 23, 27
    for design professional, 16, 20
    need for, 20
    team from design professional, 20
Inspection system:
    basic concern of, 55
    beneficiaries of, xi
    check for compliance, xi
    as checks and balances, xii, 40, 45, 53
    construction, xi
    documents, 78–166
    groups as part of, 41, 53
    joint inspection in, 53
    list of inspections, 99
    need for, xi, 55, 65
    participants in, xi
    planning for, 40
    as public protection, xi, 53, 55
    purpose/aim of, xi, 40, 53
    quality control in, 42
    recommendations for improving, 75
    safeguards in, xi, xii
    safety factors, xi
    see also, Building inspection; Construction inspection
Interest:
    of building inspector, 55
    conflicting:
        design professional as construction manager, 20, 27
        design professional as moderator, 16
        owner-design professional, 9
    in cost control, 31, 34
    decision forced, 33, 51
    financial with project, 30, 34, 49, 51
    owner's versus public, 53
    participants' self-, 30, 35, 45, 49, 55

Job history:
    in court, 66, 67
    example of forms for, 77–116
    governmental field records for, 59
    need for, 59, 67, 77
    as records, 59, 67
Joint inspections:
    governmental, 53
    other groups form, 41

Lives lost on fires, 68–70, 72
Local control of codes, 75

Misunderstandings, resolution of, 5

Observation during construction:
   for quality control, 42
   scheduled visits for, 10, 16
   timely, 16, 34
Organization:
   of building department, 59
   of contractor, 35, 36, 45
   for inspection by design professional, 13, 14, 20
   of project enhanced by inspection, xii, 36
   of subcontractor, 46, 49
Owners:
   attitude of, 9, 30, 58
   basic interest, 30
   in communications, 31, 33
   interest *vs.* public, 53
   as party to contracts, 31, 47–50
   relationship:
      to design professional, 30
      to other participants, 32, 34
      representation, 31, 33

Plan review, before permit issuance, 55
   checklist of discrepancies, 95–96
   function of permit/inspection system, 55, 59
   purpose of, 59
Plans examiners, 59, 67
Productivity:
   limits on increased, 1
   scheduling for, 1, 36, 44
Project atmosphere:
   factors contributing to, 3
   importance of, 2

Quality as cost factor, 4
   control, 5, 42
   within design concept, 4
   material, 4
   minimum standards, 4

   role of inspection system, 4, 41
   workmanship, 4, 41, 42

Regional variations:
   in attitudes, 4
   influences, 4
   in manpower availability, 4
   in types of construction, 4
Re-inspection:
   not feasible, II, 33, 62
   impedes progress, 34
   to remedy vandalism/deterioration, 33
Responsibilities:
   of building inspector, 53, 55, 59, 64
   of contractor, 35
   coordination of, 35
   of design professional, 58
   implied obligations and, 35
   of owner, 58
   of subcontractor, 46

Specifications as function of code/permit system, 53, 55
Subcontractor:
   contractual requirements, 46
   limits of liability for, 51
   as part of inspection system, 47
   responsibilities of, 46

Team:
   ability of, 3
   attitude of, xii, 9
   characteristics of, 3
   concept of inspection by, 41, 53
   effort, xii
   inspection by design professional, 20
   owner and, 31, 32

Workmen, attitude of, 1, 43, 45, 51, 52
Work sequence, scheduling of, 36, 44